Rural Development through Carbon Finance

DEVELOPMENT ECONOMICS AND POLICY

Series edited by Franz Heidhues, Joachim von Braun and Manfred Zeller

Vol. 62

PETER LANG

Frankfurt am Main · Berlin · Bern · Bruxelles · New York · Oxford · Wien

Rural Development through Carbon Finance

Forestry Projects under the Clean Development Mechanism of the Kyoto Protocol

Assessing Smallholder Participation by Structural Equation Modeling

Sebastian M. Scholz

PETER LANG
Internationaler Verlag der Wissenschaften

Bibliographic Information published by the Deutsche Nationalbibliothek
The Deutsche Nationalbibliothek lists this publication in the Deutsche Nationalbibliografie; detailed bibliographic data is available in the internet at <http://www.d-nb.de>.

Zugl.: Gießen, Univ., Diss., 2008

This dissertation was submitted to and accepted by Department 09 (Agricultural Sciences, Ecotrophology, and Environmental Management) of the Justus-Liebig-University Gießen, Germany.
Research for this dissertation was conducted at Bonn International Graduate School for Development Research being part of the Center for Development Research (ZEF) in Bonn, Germany, and funded by a PhD research grant of the Robert Bosch Foundation ('International Aspects in Agricultural & Forest Sciences').

D 26
ISSN 0943-1338
ISBN 978-3-631-59250-2

www.peterlang.de

"When we plant trees, we plant the seeds of peace and seeds of hope.
We also secure the future for our children."

Prof. Wangari Maathai
Nobel Peace Prize Laureate 2004

Preface

In its recent Fourth Assessment Report, the Intergovernmental Panel on Climate Change (IPCC) calculates that about 20 percent of anthropogenic carbon dioxide emissions result from land use change and primarily deforestation. At the same time, it is estimated that roughly one fourth of total emissions have been absorbed by terrestrial ecosystems through re-growth on cleared land, land management practices and fertilizing effects of carbon dioxide and elevated nitrogen deposition. Hence, forests are an integral part of the global carbon cycle and play an important role for the global climate, in addition to their significance for biodiversity and products that are crucial for local people as well as for society as a whole.

The international negotiations under the United Nations Framework Convention on Climate Change (UNFCCC) just lately added authority to the climate-relevance of forest ecosystems by adopting the 'Bali Action Plan'. This framework addresses enhanced national and international action on mitigation of climate change, including the "consideration of policy approaches and positive incentives (...) to reducing emissions from deforestation and forest degradation in developing countries, and the role of conservation, sustainable management of forests and enhancement of forest carbon stocks in developing countries." It is clear that forests will play a more prominent role in the post-2012 climate regime, which is highly desirable for a more holistic and concerted approach to tackle climate change in the post-Kyoto era.

But even before considering new mechanisms and approaches it is equally important to scrutinize the already existing portfolio of options under the Kyoto Protocol and to analyze if and how those mechanisms have met the expectations placed in them. To this end the study at hand delivers a valuable and timely contribution to the international discussion. It embarks on the assessment of the so-called Clean Development Mechanism (CDM) as a means to promote rural development and to simultaneously contribute to climate protection. This twin objective is the rationale of the CDM, one of the flexible implementation instruments of the Kyoto Protocol. The idea of the CDM is to meet the emission reduction commitments of industrialized countries under the strict condition of economic efficiency while promoting sustainable development in the non-industrialized world. The study hypothesizes that CDM projects in the land use and forestry sector are an efficient instrument of climate change mitigation that contributes to rural development and poverty alleviation at the same time. The potential to do so is high since roughly two thirds of the world's poor live in rural areas, with agriculture and forestry being the major source of their subsistence. Therefore, CDM operations in this sector are envisaged to yield a double dividend. However, the study also cautions that practical advice for project development is very much needed in order to entirely explore the capability of the current CDM regulations for rural development. In this context, the report postulates that the design and implementation of projects in the land use and forestry sector must be based on a thorough understanding of household and community activities and decisions if local actors are to benefit from the climate laws and regulations agreed upon internationally.

To this end, the study analyzes socio-economic aspects of a forestry project established under the CDM rules in more detail and considers an East African case study exemplarily. An agricultural household survey in Tanzania delivered the empirical household and community data for the structural equation model (SEM), which is at the center of the analysis. This innovative modeling approach is a combination of regression, path, and factor analytic models. Structural Equation Modeling aims to analyze a hypothesized context more exhaustively by taking directly observed variables as expression of an underlying, so-called 'latent' root cause to be further explored. Here the intention was to reveal underlying causalities determining the rural household's CDM participation decision. A major finding of the study in this regard is that social capital or the 'rootedness' of the farm household within the community, is much more important for a participation decision than previously thought. The study shows that social networks play a major role for adoption decisions in rural Tanzania. A fact that might be easily overlooked since common agricultural adoption studies most often rather focus on aspects related to farm resource endowment.

Furthermore, the study makes clear that apparent synergies between natural resource management and rural development do not emerge all of a sudden just because both goals are appealing. Taking the design of the carbon finance payment mechanism of the CDM tree planting activity in Tanzania as an example, it is shown that a thoughtfully designed institutional structure is essential to foster both, rural livelihoods and natural resource management. This must take new aspects like mobility considerations of farm households in rural areas into account.

With regard to the poverty-relevant aspect of the tree planting activity, an important finding is that the program design can promote farmers with constraints on land availability to actively participate. The major reason for the latter finding was the Small Group structure of the program. The Small Groups enabled farmers to plant trees collectively. The carbon finance revenues generated through the activity were then shared according to plan within the Small Group - a very relevant finding for a design that seeks to avoid discrimination of relatively resource poorer farmers.

Looking at different farm assets the study also shows that the benefits of land use-related climate projects go way beyond pure mitigation but also have a positive impact on a very broad asset base on which poor farm households depend. These positive effects contribute to the adaptive capacity of the household since natural assets, social assets, human assets, physical assets, and political assets are strengthened.

In its concluding argumentation the analysis emphasizes that the current CDM only allowing for afforestation and reforestation projects in developing countries is too restricted. Current rules only recognize environmental services related to active storage of carbon in newly established trees, while carbon conservation, for example the protection of native forests, is not rewarded as a climate-relevant service provision. In order to more holistically recognize the role that forests, alongside agriculture and other land uses play to mitigate climate change, a post-2012 climate agreement ought to expand the scope for financial incentives to emission reductions from all land-use and forest-related changes in carbon stocks. Developing countries today administer the majority of the world's environmental resources, such as tropical forests, and in doing so provide a vital global public good. They will have to be integrated into a more comprehensive incentive framework that also rewards forest conservation and sustainable forest management while promoting sustainable rural development.

Fortunately, the very recent climate negotiations in Bali and Poznań indicate a forward-looking attitude toward mitigation options from forestry and agriculture that show substantial developmental benefits for rural communities, on the one hand, and provide a flexible and supplementary mitigation solution for industrialized countries, on the other.

In this regard the study at hand is a piece of research and demonstration regarding project management, benefit distribution, and stakeholder engagement that ought to inform the political process struggling for a climate agreement to follow Kyoto.

Prof. Dr. Franz Heidhues
Center for Tropical Agriculture
University of Hohenheim, Germany

Prof. Dr. Joachim von Braun
International Food Policy Research Institute (IFPRI)
Washington D.C., USA

Prof. Dr. Manfred Zeller
Institute of Agricultural Economics and Social Sciences in the Tropics
and Subtropics, University of Hohenheim, Germany

Table of Contents

List of Tables

List of Figures and Boxes

List of Acronyms and Abbreviations

ADF	Asymptotically Distribution-Free
AGFI	Adjusted Goodness-of-Fit Index
AIC	Akaike Information Criterion
A/R	Afforestation and Reforestation
BIOCF	BioCarbon Fund
CAAC	Clean Air Action Corporation
CAIC	Consistent Akaike Information Criterion
CDM	Clean Development Mechanism
CERs	Certified Emission Reductions
tCERs	Temporary Certified Emission Reductions
lCERs	Long-term Certified Emission Reductions
CFI	Comparative Fit Index
CH_4	Methane
CMIN	Chi-square
CO_2	Carbon Dioxide
COP	Conference of the Parties
C.R.	Critical Ratio
DF	Degrees of Freedom
DIAS	District Intergraded Agricultural Survey
DNAs	Designated National Authorities
DOEs	Designated Operational Entities
EB	Executive Board
ECVI	Expected Cross Validation Index
EPC	Expected Parameter Change
EU	European Union
FAO	Food and Agriculture Organization of the United Nations
GFI	Goodness-of-Fit-Index
GHG	Greenhouse Gas
GLS	Generalized Least-Squares
GPS	Global Position System
GWP	Global Warming Potential
HFC	Hydrofluorocarbon
HH	Household
ID	Identification
JI	Joint Implementation
LDCs	Least Developed Countries
L-EX	latent exogenous

L-EN	latent endogenous
LSMS	Living Standards Measurement Study
LULUCF	Land use, Land-use Change and Forestry
M-EN	measured endogenous
MIs	Modification Indices
ML	Maximum-Likelihood
MOP	Meeting of the Parties
NASC	National Agriculture Sample Census
NCP	Noncentrality Parameter
NFI	Normed Fit Index
NGO	Non-governmental Organization
N_2O	Nitrous oxide
NPAR	Number of Parameters
NTFP	Non-timber Forest Products
ODA	Official Development Assistance
PCFI	Parsimony Comparative Fit Index
PCLOSE	Closeness of Fit
PDA	Personal Digital Assistant
PDD	Project Design Document
PEN	Poverty Environment Network
PES	Payments for Environmental Services
PFC	Perfluorocarbon
PGFI	Parsimony Goodness-of-Fit Index
PNFI	Parsimony-Adjusted Normed Fit Index
PRATIO	Parsimony-Ratio Index
REDD	Reduced Emissions from Deforestation and Degradation
RMR	Root Mean Square Residual
RMSEA	Root Mean Square Error of Approximation
SACCOS	Savings and Credit Cooperatives
SEM	Structural Equation Model
SF_6	Sulphur hexafluoride
SLA	Sustainable Livelihood Approach
SLS	Scale Free Least-Squares
TIST	The International Small Group & Tree Planting Program
TLI	Tucker-Lewis Index
TZS	Tanzanian Shillings
ULS	Unweighted Least-Squares
UMET	Ukuzaji Maendeleo Endelevu Tanzania

UN	United Nations
UNFCCC	United Nations Framework Convention on Climate Change
US	United States of America
WHO	World Trade Organization
ZEF	Center for Development Research

Acknowledgements

The volume at hand is the result of my doctoral research project at the Bonn International Graduate School for Development Research, which is part of the Center for Development Research (ZEF) of the University of Bonn. The project would not have been possible without the generous academic as well as practical support of a number of persons and institutions.

First and foremost, I would like to express my gratitude to the German Robert Bosch Foundation, which has supported my work with a dissertation research fellowship in the program area "Sustainable Use of Natural Resources".

Certainly, the inspiring work environment at the Center for Development Research has academically fostered my work. Here, I would like to thank my former tutor Dr. Bettina Hedden-Dunkhorst and my ZEF colleague and friend Dr. Jan Börner for their motivating ideas during the first stage of this research. I am also thankful to Prof. Dr. Arild Angelsen who has reviewed a very first proposal of my work and encouraged me a lot through his incredible effort to build the Poverty Environment Network (PEN), an international network and research project on poverty, environment and forest resources.

Methodologically, I benefited very much from the discussions with Dr. Christian Rietz from the Center for Evaluation and Methods of the Rhenish Friedrich-Wilhelm University, Bonn.

The field research for this dissertation in Tanzania was only possible through the assistance of Dr. John M. Msuya, also a ZEF graduate, now teaching at Sokoine University of Agriculture, in Morogoro, Tanzania. In addition, I would like to mention the graduate students from Sokoine University who traveled with me through rural areas and facilitated the interviews with the farm households. I am thankful to John Jeckoniah, Victor George, July Zablon, Johari Massengi, and Lulu Luflenge.

I equally feel obliged to the persons working for TIST – The International Small Group and Tree Planting Program. Ben Henneke, Charles E. Williams, Ramesh Narasimhan, Sikitu Frank, and all staff of Ukuzaji Maendeleo Endelevu Tanzania (UMET) gave me generous insights into their often challenging work. The very first idea to do research on an African forest-related project under the Clean Development Mechanism came from my World Bank colleague Benoît Bosquet, PhD, whose advice is very much appreciated.

Last but not least, I am indebted to my doctoral advisors Prof. Dr. Ernst-August Nuppenau and Prof. Dr. Stefanie Engel who supported me at all times throughout this undertaking.

1. Introduction

Development of rural areas and protecting the global climate at the same time: Is that possible? Too good to be true, one might think. Yet, this twin objective, in essence, describes what the so-called Clean Development Mechanism in the area of land use-related project activities is envisaging. But let us start chronologically.

In 1997, when the international community agreed on the Kyoto Protocol as the first-ever international, legally binding treaty to counteract global warming, the signatory countries assured themselves with a certain degree of flexibility to meet their pledged emission reduction commitments. The idea of the so-called 'flexible mechanisms' of the Kyoto Protocol was created. The underlying concept of those mechanisms is to meet the Kyoto commitments under the strict condition of economic efficiency. In effect, the flexible mechanisms aim to reduce emissions from the source with the lowest marginal costs of abatement. The costs of reducing emissions can vary substantially from region to region and technology to technology (SANDS (2004)). However, the benefit for the atmosphere is the same, presumably, irrespective of the fact where action is taken. One of the three mechanisms guaranteeing compliance flexibility is the abovementioned Clean Development Mechanism (or CDM in short).

The CDM provides a framework for investments in greenhouse gas (GHG) mitigation projects in developing countries. In doing so, the goal of the CDM is two-fold:

1. Developing countries, the so-called non-Annex I countries, are supposed to benefit from project activities fostering sustainable development while at the same time resulting in certified emission reductions (CERs) through the project activities.[1]
2. Industrialized countries, investing in such projects, can use the accruing CERs to contribute to compliance with at least part of their quantified emission limitation and reduction commitments under the Kyoto Protocol.

The numerous opportunities for mitigating atmospheric carbon emissions in developing countries include projects in the land use, land-use change and forestry (LULUCF) sector, often also called 'sinks' projects. In theory, such projects can include afforesting or reforesting degraded lands, implementing sustainable agricultural practices leading to increased soil carbon levels, or even slowing tropical deforestation. All these projects have in common that either the sequestration (i.e., storage) of carbon in biomass is enhanced or the release of carbon from biomass should be prevented. LULUCF projects take the sequestration of carbon in biomass into account and accordingly generate marketable emission reduction units. Thus, carbon becomes a tradable commodity. However, for po-

---

1 Developing countries are not listed in Annex I of the United Nations Framework Convention on Climate Change (UNFCCC) and are therefore called 'non Annex I countries' in the jargon of the international climate negotiations.

litical reasons, the Parties of the United Nations Framework Convention on Climate Change (UNFCCC) have limited the use of land use-related project categories in the CDM considerably. Only 1% of Annex I countries' Kyoto obligations can be met through those projects. Furthermore, eligible project activities are restricted to 'afforestation' and 'reforestation' activities for the time being.

However, this limitation of land use-related CDM projects is clearly at odds with findings regarding rural development and poverty alleviation. Forestry projects are considered a low cost and rapidly implementable means of climate change mitigation that can additionally contribute to poverty alleviation. Since approx. 70% of the world's poor are located in rural areas, with agriculture and forestry being a major source of their subsistence, a double dividend is expected (SUNDERLIN et al. (2005); THE WORLD BANK (2005)). The CDM in the LULUCF sector is expected to show substantial development benefits for project proponents, i.e., smallholders and rural communities in developing countries (NILES et al. (2002); NILES et al. (2002)). It is hypothesized that, in particular, land use-related projects are a direct and feasible investment at the local level which does not require a complex capacity building up front like other CDM project categories, e.g., in the energy sector. It is suggested that agroforestry projects can provide livelihoods, foster deforestation prevention and conserve irreplaceable biodiversity, notably, if it is part of an integrated approach that reduces pressure for forest conversion elsewhere. And on top, such projects should provide a continuous revenue stream by the sale of carbon certificates and can therefore make a direct pecuniary contribution to the alleviation of rural poverty.

Studies showed that particularly for Africa LULUCF projects are a major comparative advantage in the emerging international carbon market (HAMILTON et al. (2007)). At present, Africa has only very few carbon credits from other project categories to offer on the carbon market. Since the Kyoto Protocol does not address forest conservation or the prevention of deforestation, developing countries are restricted in their opportunities to benefit from the CDM (STRECK and SCHOLZ (2006)). A reasonable and comprehensive post-Kyoto regime will have to expand the existing system by creating a framework encompassing all LULUCF-related project categories. However, for the time being, concrete and policy-oriented advice is needed in order to explore the entire potential of the current CDM regulations for rural development. If local actors are to benefit from the set of laws agreed upon internationally, we hypothesize that the design and the implementation of A/R projects must be based on a sound understanding of household and community activities and decisions.

To this end, the study at hand analyzes socio-economic aspects of A/R projects established under the rules of the international climate regime in more detail and considers an African A/R project exemplarily.

As will be shown, field research in Tanzania delivered empirical and first-hand, household and community-based data to run a structural equation model. Structural equation models (SEMs) combine various types of models known from other contexts, such as regression, path, and factor analytic models. SEMs aim to analyze a hypothesized context more exhaustively by taking observed or measured variables as expression of an underlying root cause to be explored and thus look for underlying causalities. Thus, not only directly observable, but also hypothesized 'latent' variables can be modeled in structural equation models, which allows for a more profound and holistic analysis, not only of the collected data, but also of the situation on the ground.

1.1 Objectives of the Study

This study focuses on the nexus of climate protection and its potential for development of rural areas through CDM projects in the land use and forestry sector. Given the fact that up to this date almost no studies exist with regard to the local effects of land use-related CDM projects, the first objective of the study is to

I. contribute to the understanding of the inter-linkages between international climate policy and national development policy through the CDM.

In consideration of the fact that the local impact of international decisions under the UNFCCC is of importance for potential development on the ground, the second objective is to

II. put the rural household at the center of analysis and to assess factors that determine the participation of smallholders in LULUCF carbon finance projects.

The international climate regime, its rules, and regulations are still evolving rapidly. So far, the huge majority of the internationally agreed rules is only valid until 2012, i.e. until the end of the so-called 'first commitment period' of the Kyoto Protocol. Therefore, the third objective of the study is to

III. give a rather long-term outlook with regard to the socio-economic impacts, LULUCF projects have on subsistence farmers.

The aforementioned objectives are of interest because the rules of the future CDM not necessarily go on with the existing framework but are going to be revisited before the end of 2012. Thus, there is a window of opportunity to stress positive aspects of the existing regulations leading to desired livelihood impacts. In the same way, potential adverse impacts can be reconsidered in the international decision-making process and altered accordingly. The fourth objective is hence to

IV. provide policy-oriented advice with regard to the local design and development of future LULUCF projects under the CDM.

1.2 Data, Methodology, and Hypotheses

Data for the analyses of this study was collected on the ground in Tanzania. As will be described further down, a field research campaign with agricultural household surveys implemented the household-centered research approach of this study.

Based on the aforementioned objectives, the study aims to assess the following hypotheses in more detail. Firstly, we put forward that

i. thoughtfully designed LULUCF projects under the Clean Development Mechanism of the Kyoto Protocol are able to combine both, aspects of climate change mitigation and rural development.

With regard to the methodology applied, the study hypothesizes that

ii. in order to contextually assess factors that determine the participation of smallholders in LULUCF carbon finance projects, structural equation modeling (SEM) offers the most suitable approach. SEM is not only confined to measure or observe variables but ideal to reveal underlying root causes of an adoption decision by modeling latent variables.

iii. Furthermore, it is assumed that land use-related climate mitigation projects go way beyond in their influence and have a positive impact on a very broad asset base on which poor farmers depend. This includes, but is not limited to natural assets, social assets, human assets, etc.

The study also addresses a major shortcoming of common agricultural adoption studies. The latter often mainly focus on issues related to farm resource endowment whereas aspects of social capital, i.e., social networks in rural areas that play a major role for adoption decisions are often overlooked. This results in the fact that

iv. the study at hand hypothesizes that particularly social interactions, the connectedness of the farm household within the village, is of utmost importance for the participation decision in land use related project categories of the CDM.

1.3 Content Outline

After this brief introduction, giving a first idea of the general topic, the objectives, and main hypotheses the dissertation is going to deal with, the subsequent chapters are structured as follows: Chapter 2 provides background information on the international climate regime, its flexible mechanisms and particularly the CDM. Special emphasis lies on the rules for land use, land-use change and forestry (LULUCF) projects as related to carbon sequestration. LULUCF projects are examined with regard to their potential development benefits. To do so, the CDM in the forestry sector is treated as a special case within the broader framework of payments for environmental services (PES). While PES were primarily designed as a mechanism to improve the efficiency of natural resource management, possible synergies with poverty reduction and rural development are discussed in the literature. However, concerns about potential adverse impacts of PES payments exist, too. Chapter 2 outlines the current knowledge in detail and discusses the applicability for the CDM.

Chapter 3 describes the case study setting, field research, and the agricultural household survey which was applied. The modeling undertaken in the subsequent chapters is based on the data gathered during field research in Tanzania. In addition, chapter 3 presents a general overview of the 'International Small Group & Tree Planting Program' (TIST), which is locally involved in a carbon-related PES scheme. The agricultural household survey, its sampling procedure, and the data collection in Tanzania are described in detail before case study results are presented by means of mainly descriptive statistics. Special emphasis lies on socio-economic characteristics of the interviewed households, descriptives enlightening the farming system and the livelihood strategies of the interviewees.

The objective of chapter 4 is to assess in more detail under which conditions local farmers are in a position to participate in carbon finance projects. As will be shown, PES as well as forestry-related projects in the CDM are too recent for a straightforward assessment of participation decisions. Thus, scientific literature on participation in other rural programs or on the introduction of new agricultural technologies has been consulted in order to provide insights on parallels. At first, the results of a broader literature review on agricultural adoption are presented. Subsequently, chapter 4 develops a more specific theoretical model for the adoption of carbon farming. At the end of the chapter, particular factor categories determining participation are hypothesized. Those factor categories already introduce the subject of the subsequent chapter 5, which is structural equation modeling.

In chapter 5, a full structural equation model is built, estimated, tested, and modified. The chapter introduces the general idea of structural equation modeling and at first formulates an initial model based on the outcomes of chapter 4. In this initial model the identified factor categories potentially explaining adoption are brought together to form a first structural equation model. Subsequently, the model results are discussed in detail and the initial model is revised to ensure better fit and greater explanatory power. The 'lessons learned' of the modeling process directly feed into the synthesis of chapter 6.

Based on the modeling in chapter 5, chapter 6 elaborates on actual and potential project participants characteristics and describes their identity. Obstacles to participation are outlined and potential impacts on project participants and other groups are discussed. The asset-based approach of livelihood evaluation outlined in the last part of chapter 2 is revisited. On the basis of the modeling results we evaluate and discuss welfare gains and potential drawbacks for the assets on which poor farmers depend.

The conclusions and policy recommendations of chapter 7 finally summarize the findings of the preceding chapters. Prospective guidance with regard to the design and development of future LULUCF projects under the CDM is provided. The main findings of the dissertation cannot be summarized without pointing to the existing limitations of the findings and constraints of the methodology applied.

2. The International Climate Regime, its Clean Development Mechanism (CDM) and the Link to Rural Development

The following chapter introduces the international climate regime and particularly one of its so-called 'flexible mechanisms' for the fulfillment of emission reduction commitments, namely the Clean Development Mechanism (CDM). Emphasis is put on land use, land-use change and forestry (LULUCF) projects for carbon sequestration (storage) under the CDM framework. Subsequently, LULUCF projects are examined with regard to their potential developmental benefits, especially in rural areas. Furthermore, LULUCF projects in the CDM are considered as a special case within the broader framework of payments for environmental services (PES). While the PES approach was primarily designed as a mechanism to improve the efficiency of natural resource management, potential synergies with poverty reduction and rural development are important to look at (PAGIOLA et al. (2005)). The CDM, in contrary, explicitly establishes a twin objective of climate protection and sustainable development. Thus, the CDM is a 'special case' of PES with its own specific international regulatory framework.

2.1. The International Climate Regime and the Clean Development Mechanism (CDM)

This section briefly outlines the international climate regime consisting of the United Nations Framework Convention on Climate Change (UNFCCC) and the Kyoto Protocol. Furthermore, one of the three so called 'flexible mechanisms' of the Kyoto Protocol for the fulfillment of emission reduction commitments of industrialized countries is explained in more detail.

The international climate regime was established in June 1992 when the UNFCCC was opened for signature at the Earth Summit in Rio de Janeiro. Within one year 165 countries and the European Union (EU) agreed to sign the contract which finally entered into force on March 21, 1994 (OBERTHÜR and OTT (1999)). Today, the UNFCCC is almost unanimously accepted as the basis of international politics on climate change. Even though it was widely acknowledged that rising concentrations of atmospheric carbon dioxide (CO_2) were driving changes in climate patterns and that industrialization and economic development have resulted in a sharp rise of CO_2 and other greenhouse gas (GHG) emissions in the atmosphere, the UNFCCC did not define legally binding reduction targets for those gases. A rather 'soft' approach was adopted in order to bring as many countries as possible on board of the Convention. Still, the UNFCCC formulated 'the ultimate objective' (Art. 2, UNFCCC) of confronting the trend of rising GHG concentrations in the atmosphere and ultimately reversing it (UNFCCC (1992)). The other major obligations of the Convention outline that all Parties shall:

- develop national inventories of anthropogenic GHG emissions and removals by sinks[2],
- elaborate and implement national and regional programs containing measures to mitigate and facilitate adaptation to climate change,
- promote sustainable management of sinks and reservoirs of GHGs,
- cooperate in adaptation measures,
- promote and cooperate in the integration of climate policy considerations into other policy arenas and international cooperation as well as science technology, education etc., and
- report on inventories and relevant policies and measures (UNFCCC (1992); OBERTHÜR and OTT (1999)).

In addition, the Convention establishes institutions and procedures and hence a negotiation platform. For instance, the institutional framework and the technical infrastructure necessary to define future international solutions to climate change are also set up. The most important institution established by the UNFCCC is the 'Conference of the Parties' (COP) as the highest decision-making body of the regime (Art. 7, UNFCCC). Furthermore, in Article 3 the UNFCCC states that 'common but differentiated responsibilities and respective capabilities' of industrialized and developing countries exist with regard to mitigation of and adaptation to climate change. "*Accordingly, the developed country Parties should take the lead in combating climate change and the adverse effects thereof*" (Art. 3, UNFCCC). The Convention lists those 'developed' or industrialized countries in Annex I of the Convention. This is why industrialized countries are often referred to as 'Annex I countries' while developing countries are hence 'non-Annex I countries'. Ultimately, the UNFCCC only obliges Annex I countries to 'aim' at reducing their GHG emissions to 1990 levels by the year 2000 (Art. 4.2(b), UNFCCC).

In accordance with other international environmental treaties, the UNFCCC was further developed and broken down for implementation through the Kyoto Protocol in 1997.[3] In the Kyoto Protocol industrialized countries jointly commit themselves to reduce their aggregate GHG emissions by at least 5% below 1990 levels in the so called 'commitment period' which runs from 2008 to 2012.[4] Par-

2 The UNFCCC defines sink as 'any process, activity or mechanism, which removes a greenhouse gas, an aerosol or a precursor of a greenhouse gas from the atmosphere' (Art. 1.8, UNFCCC).

3 The 'Montreal Protocol on Substances That Deplete the Ozone Layer' from 1987, an international milestone to protect the stratospheric ozone layer for example fleshes out the 'The Vienna Convention for the Protection of the Ozone Layer' from 1985. Accordingly, the 1979 'Convention on Long-range Transboundary Air Pollution' has been extended by eight protocols that identify specific measures to be taken by Parties to cut their emissions of air pollutants.

4 The five years spanning commitment period is supposed to mitigate temporal risk, a fixed year commitment date might exhibit due to fluctuations in economic performance or extreme weather conditions. Thus, the commitment period provides the Parties with temporal flexibility to meet their commitments.

ticular and country-specific 'quantified emission limitation and reduction commitments' were outlined in Article 3.7 of the Kyoto Protocol and its Annex B (UN (1997)). The targets do not only refer to CO_2, but cover a 'basket' of six climate change-related gases, namely CO_2, CH_4 (Methane), N_2O (Nitrous oxide), HFCs (Hydrofluorocarbons), PFCs (Perfluorocarbons), and SF_6 (Sulphur hexafluoride). Those gases have to be reduced in several different 'sectors/source categories' contained in Annex A of the Protocol.[5] Thus, in contrary to the UNFCCC, the Kyoto Protocol establishes binding targets for industrialized countries and a timetable for the fulfillment of those targets. To meet their agreed goals the Parties have different options. The most obvious one is to reduce emissions at home, i.e., cut emissions at their source. This approach is commonly referred to as 'domestic action.'

Furthermore, the so-called 'flexible mechanisms' of the Kyoto Protocol provide Parties with economic and geographical flexibility to meet their commitments. The common aim of the three flexible mechanisms is to reduce the costs of complying with Kyoto targets by allowing measures to limit GHG emissions, or measures to enhance removals of GHGs from the atmosphere by sinks to be implemented where they are cheapest. While the costs of limiting emissions or expanding removals regionally vary to a large extent, the positive atmospheric effect is the same regardless where the action is undertaken (UNFCCC (2005a)). The flexible mechanisms encompass the Joint Implementation (in short 'JI', Art. 6, Kyoto Protocol) of climate protection projects between industrialized countries (i.e., two Annex I countries implement a project jointly). In addition, such projects can be implemented between an industrialized (investing country) and a developing country (host country). The latter describes the framework of the Clean Development Mechanism (CDM), outlined in Article 12 of the Kyoto Protocol.

Furthermore, the Protocol establishes a scheme for the trade with emission certificates (Emissions Trading) between industrialized countries as described in Article 17, which is the third flexible mechanism. While in Emissions Trading a country sells or buys emission certificates on the international market, the project-based mechanisms Joint Implementation and Clean Development Mechanism create new certificates for the investing Annex I country. The respective host country, either also an Annex I country for JI or a developing country in the case of CDM projects, is supposed to benefit through the investment and the transfer of new technologies and know-how. There is no quantitative limit on the extent to which the flexible mechanisms can be used to meet commitments. However, Annex I Parties are obliged to provide information showing that their use of the mechanisms is

5 In order to compare the atmospheric impact of the different gases, the IPCC (Intergovernmental Panel on Climate Change) has developed the 'Global Warming Potential' (GWP) as an index that measures the relative potential of the six greenhouse gases to contribute to global warming. To this end, the impact of all other GHGs is compared with the impact of CO_2 and referred to in terms of CO_2 equivalent (CO_2e), i.e., carbon dioxide has been designated a GWP of 1, Methane has a GWP of 23 and the other gases have even higher GWPs (see IPCC (2001)).

'supplemental to domestic action' and domestic policies and measures must constitute 'a significant element' of efforts to meet emission reduction commitments (UNFCCC (2002)).

The focus in the following is on the CDM which is detailed in Article 12.2 of the Kyoto Protocol along these lines: "*The purpose of the clean development mechanism shall be to assist Parties not included in Annex I in achieving sustainable development and in contributing to the ultimate objective of the Convention, and to assist Parties included in Annex I in achieving compliance with their quantified emission limitation and reduction commitments under Article 3*" (UN (1997)). The wording of the CDM definition establishes an unmistakable twin objective: First, to contribute to sustainable development of the host country (i.e., the developing or non-Annex I country) and second, to assist Annex I countries in achieving their reduction targets. Both, public and private entities may participate in CDM projects (Art. 12.3(a), Kyoto Protocol), which must produce measurable long-term benefits in the form of emission reductions or removals (Art. 12.5(b)), and must be 'additional' to any that would have otherwise occurred (Art.12.5(c)). The additionality clause of the CDM does not only mean that public funding for the CDM must not result in a diversion of official development assistance (ODA) but points to the crucial fact that the CDM is in principle a *carbon-neutral* process. This is because the Certified Emission Reduction units (CERs) generated through a CDM project in a developing country are transferred to the industrialized country, which is now allowed to emit an equal amount of carbon while meeting its targets. Thus, if the respective project would have taken place without the CDM transaction anyway, GHGs in the atmosphere will increase as a result of the transfer of CERs. While the Kyoto Protocol establishes the general framework of the CDM, its particular rules were laid down in the so-called 'Marrakesh Accords' in 2001. The focus of the Marrakesh Accords is on projects that reduce emissions. Rules for land use, land-use change and forestry (LULUCF) related CDM project categories that enhance the removal of GHGs from the atmosphere have since been developed quite a bit. Those rules are the focus of the following chapter.

Also, Article 12 of the Kyoto Protocol points out that CDM projects must have the approval of all Parties involved. This project approval must be obtained from designated national authorities (DNAs) to be set up by Annex I and non-Annex I Parties. The CDM then follows a project cycle, which is supervised by the CDM Executive Board (Art. 12.5 and 4, Kyoto Protocol). In addition, the Executive Board is guiding and overseeing all practical arrangements of the CDM. Moreover, all CDM projects must be based on appropriate, transparent, and conservative baselines[6] and must have in place a precise monitoring plan to collect accurate

6 A baseline is the starting point for measuring emission reductions or removals. A baseline must be devised according to a methodology that has been approved by the CDM Executive Board. In principle, project participants can develop a new methodology for their respective project, but every methodology must first be authorized and registered by the Executive Board.

emissions data. Independent organizations, so-called 'designated operational entities' (DOEs), also play a key role in the CDM project cycle. Their task is to validate and certify CDM projects according to the rules developed by the international community. Table 2-1 outlines the full CDM project cycle.

Table 2-1 The CDM Project Cycle

Step	Actor	Explanation
1.Project Design	Project Proponents	Proponents must prepare a project design document (PDD), including a description of the baseline and monitoring methodology to be used, an analysis of environmental impacts, comments received from local stakeholders and a description of new and additional environmental benefits that the project intends to generate.
2.Validation and Registration of the Project	Designated Operational Entity (DOE) and Executive Board (EB) of the CDM	An operational entity will then review the aforementioned documentation of the project proponents and, after providing an opportunity for public comments, the DOE decides whether or not to validate the project. When a project is duly validated, the DOE will forward it to the Executive Board for formal registration.
3.Monitoring	Project Proponents	Once a project is running, the proponents will monitor it. They will prepare a monitoring report, including an estimate of CERs generated by the project, and will submit it for verification by an operational entity.
4.Verification / Certification	DOE	To avoid conflict of interest, the verification / certification of a CDM project must be carried out by a different operational entity to that which validated the project in the beginning (except for so-called small scale CDM projects). Following a detailed review of the project, which may include an onsite inspection, the operational entity will produce a verification report and ultimately certify the occurred emission reductions as 'real'.
5.Issuance of CERs	EB	Unless a participating Party or three Executive Board members request a review within 15 days, the Board will issue the CERs and distribute them to project participants / proponents as requested.

Source: UNFCCC (2006b) and UNFCCC (2005a) - modified

All certified emission reductions (CERs) generated by CDM projects are furthermore subject to a levy. This 'share of the proceeds' specifies that two per cent of the CERs from each project have to be paid into an 'Adaptation Fund' set up to help particularly vulnerable developing countries to adapt to the adverse effects of climate change. Merely CDM projects in least developed countries (LDCs) are exempt from paying this share of proceeds.[7] The level of another CDM specific tax is yet to be determined: in the future, a certain percentage of CERs will be levied on projects to cover the CDM's administrative costs.

2.2. Land Use, Land-Use Change and Forestry (LULUCF) Projects under the CDM and the Generation of 'Sequestration Credits'

As outlined above, the Kyoto Protocol has established targets that can partly be met through project activities in other countries. The setting of the CDM allows an Annex I country and a non-Annex I country to jointly realize a project leading to emission reductions or enhanced removals by sinks in the non-Annex I country. Projects can be implemented in several sectors including the energy, the industrial, the agricultural, and the waste management sector. Project activities from land use, land-use change and forestry (LULUCF) can also be used to reach the targets. The CDM allows credits to be acquired not only for reduced emissions, but also for projects that help in the sequestration of CO2 (see Box 2-1). Thus, the CDM as well as the UNFCCC and the Kyoto Protocol acknowledge the vital role that land use-related activities and especially forests play for the global climate. Several articles in both, the UNFCCC and the Kyoto Protocol, refer explicitly to 'sinks'. The UNFCCC recognizes that all sinks and reservoirs of GHGs have an important impact on terrestrial and marine ecosystems (UNFCCC (1992), Preamble, Para. 4)). Among the desired means of attaining the Convention's objective of 'stabilization of greenhouse-gas concentrations' is the use of the capacity of sinks to reduce the concentration of GHGs in the atmosphere. Even prior to the UNFCCC, the fact that vegetation and soils can accumulate carbon, thereby reducing the rate of CO_2 build-up in the atmosphere that is responsible for climate change was recognized as a means for climate protection (IUCC (1993)). Ever since, this topic has figured on the agenda of international climate negotiators.

During the negotiations of the Kyoto Protocol, however, the actual knowledge of the delegates with regard to the quantification of emissions from the LULUCF sector was rather limited. Though, it was clear that the complete inclusion or exclusion of sinks would significantly influence the emission budget of the Parties. The option to account for sinks was seen by some Parties and NGOs as an unfortunate loophole to water down the emission reduction efforts of the Protocol (FEARNSIDE (2001)). For supporters, the inclusion of sinks was an economic necessity for achieving Kyoto targets, whilst others pointed to the massive quantities of carbon exchanged between the atmosphere, vegetation, and soils every year, an amount 30 times higher than the emissions from fossil fuels (WATSON and NOBLE (2004)).

7 For the criteria for the identification of the Least Developed Countries (LDCs) see UN (2004).

Box 2-1 Forests and Carbon Sequestration

Over time forests accumulate carbon through the growth of trees and the increase of soil organic carbon unless major disturbances occur (i.e., the forests 'sequester' or absorb carbon). Immature forests sequester carbon at high rates, while in mature forests carbon sequestration eventually equals decomposition; that is, the carbon balance of the ecosystem reaches a steady state. The forest is a carbon reservoir, but no longer acts as a carbon sink. This implies that whether forests act as reservoirs, active sinks for carbon from the atmosphere, or sources of GHGs depends on several factors such as the age of the forest, the management regime, other biotic and abiotic disturbances (e.g. insect pests, forest fires etc.) as well as human-induced deforestation.

They were concerned that by omitting sinks a major exchange of carbon, which could swamp any gains made through activities under the Kyoto Protocol, would be ignored. An eventful and sometimes fairly controversial negotiation process followed that ultimately led to a complicated accounting system for LULUCF (SCHULZE et al. (2002)). With regard to the CDM it took almost six years of assessment and negotiation to agree on specific LULUCF-related rules. Finally, negotiations reached consensus at the ninth session of the Conference of the Parties (COP 9) in December 2003. At this conference, an agreement on the definitions and modalities for the use of afforestation and reforestation (A/R) projects in the CDM was finalized. Prior to COP 9, Parties had decided in the Marrakesh Accords to limit the use of LULUCF in the CDM to only 'afforestation and reforestation' projects.[8] In addition, the use of forestry projects in developing countries was restricted to meet only one per cent of Annex I Parties' Kyoto obligations.[9] The major reason for the aforementioned restrictions with regard to LULUCF projects in the CDM is the issue of 'permanence' of carbon storage.

All LULUCF-related activities differ in one significant way from emission reducing activities in other sectors. Whereas a tonne of emission reductions, once achieved, remains a benefit to the atmosphere, a tonne of sequestered carbon is of benefit to the atmosphere only as long as it remains sequestered. If a tree is felled,

8 Per definition in UNFCCC (2006c), 'afforestation' is the direct human-induced conversion of land that has not been forested for a period of at least 50 years to forested land through planting, seeding and/or the human-induced promotion of natural seed sources. 'Reforestation' is the direct human-induced conversion of non-forested land to forested land through planting, seeding and/or the human-induced promotion of natural seed sources, on land that was forested but that has been converted to non-forested land. For the first commitment period, reforestation activities will be limited to reforestation occurring on those lands that did not contain forest on 31 December 1989.

9 For the most important definitions and rules on LULUCF, see 16/CMP.1 (decision 11/CP.7 before adoption by COP/MOP1), UNFCCC (2006c).

stored carbon is released and the temporary climate benefit reversed: in other words, the benefit is 'non-permanent'. The permanence problem is of specific interest as the CDM is implemented in countries without binding emission targets (i.e., in developing countries). This implies that any loss of carbon *after* the transfer of carbon credits to the industrialized country creates carbon credits that no longer refer to actually sequestered carbon, so called 'paper tonnes' (STRECK and SCHOLZ (2006)).

To counteract the problem of permanence a number of schemes based on creating credits only valid for a certain period of time, i.e., 'temporary credits' or even insurance systems were suggested. Eventually, COP 9 agreed on two forms of temporary credits for LULUCF - CDM projects, namely temporary CERs (tCERs) and long-term CERs (lCERs).[10]

Before the two forms of CERs are discussed in more detail, the length of the crediting period, i.e., the period over which the LULUCF project can generate carbon credits should be noted. In LULUCF CDM the agreement between the buyer and the seller can be extended for a crediting period of either non-renewable 30 years, or 20 years, twice renewable up to a maximum of 60 years. For both forms of certificates (tCERs and lCERs), an independent verifier, i.e., a designated operational entity (DOE), must confirm every five years that any sequestered carbon, that has been used to generate credits, remains sequestered. More precisely, the verifier confirms that the project retains at least the amount of carbon to cover the credits issued.

Hence an accounting system monitors a number of specific carbon pools within the project boundary. These include above-ground biomass, below-ground biomass, litter, dead wood and soil organic carbon. Particular carbon pools can be excluded if it can be shown that they are no source of carbon. If there has been a net loss of sequestered carbon between two 5-yearly certifications, lCERs have to be replaced within a month. The expiring tCERs have to be replaced only at the end of the commitment period. Thus, for tCERs the expiry, verification and re-issuance of tCERs correct for any losses of carbon from a project while with lCERs, if net carbon stocks decrease between two certification reports, the difference in lCERs needs to be replaced. If in contrary carbon stocks increase during two certification reports an equal amount of new lCERs may be issued by the Executive Board. In addition, COP 9 introduced a rule stating that at the end of the agreed crediting period all tCERs and lCERs must be replaced by credits from other projects. This creates a liability to replace both forms of temporary credits at some future point in

10 See decision 5/CMP.1, Annex, Section A. Definitions paragraph 1(g) "Temporary CER" or "tCER" is a CER issued for an afforestation or reforestation project activity under the CDM which (...) expires at the end of the commitment period following the one during which it was issued;

(h) "Long-term CER" or "lCER" is a CER issued for an afforestation or reforestation project activity under the CDM which (...) expires at the end of the crediting period of the afforestation or reforestation project activity sunder the CDM for which it was issued, UNFCCC (2006d).

time, depending on the end of the crediting period. While some parties feared that such a liability would deter the use of LULUCF in the CDM, others stated that liabilities at least 20, and as much as 60 years in the future would have little practical impact on the use of LULUCF in the CDM (SCHOLZ and NOBLE (2005)).

Some further aspects have to be addressed in the development of LULUCF CDM projects. In all CDM projects it must be proven that the carbon sequestration (or emission reduction) would not have occurred without the incentives provided by the existence of the Kyoto Protocol, i.e. that the project is additional. Without this test there would be no benefit to the atmosphere. The CDM Executive Board provides a step-wise tool to demonstrate additionality of prospective LULUCF projects in the CDM (UNFCCC (2005c)). The developers of A/R projects are strongly encouraged to make use of the tool, which screens a potential project for additionality in five subsequent steps.

Another issue to be addressed is that of *'leakage'* which is not unique for LULUCF projects, but plays an important role in assessing the amount of credits to claim for any CDM project. Leakage refers to an unintended change in activities or behavior outside the project area that, partially or totally, offsets the climate gains of the project. These changes may lead to reduced sequestration or increased emissions outside the project boundary, negating some of the climate benefits of the project. Critics of LULUCF projects in the CDM were concerned that it would result in widespread establishment of monoculture plantations that would drive people off their land only to deforest areas elsewhere.

Seen from an economic perspective, leakage is a biologically inspired name of general equilibrium effects (ANGELSEN (2005)). Such effects are neither intrinsically condemnable nor is it possible to completely avoid them. In reality, it might be very costly for a project to try to entirely eliminate any potential leakage. This implies that project developers have to thoroughly address leakage in the project design and account for it by subtracting potential unintended GHG emissions due to the project activity from the project performance (SCHOLZ and JUNG (2008)).

In addition, any CDM project should locally create *sustainable development benefits* beyond the global atmospheric benefits. When the Kyoto Protocol was first negotiated, some NGOs expressed concern that the CDM might unintentionally promote the development of LULUCF projects that were detrimental to local communities and the environment. Again it was feared that the CDM would foster monoculture plantations, which sequester carbon cheaply, but at the expense of sustainable livelihoods and biodiversity. However, the CDM additionality test, mentioned above, means in principle that typical large-scale timber plantations, which are economically viable without CDM-generated income, should not be eligible for crediting (UNFCCC (2005c)).

Furthermore, a project design document (PDD) must describe the project's socio-economic and environmental impacts. Individuals, groups, or communities affected by the proposed CDM project activity must be consulted. Project participants have to submit documentation on the analysis of socio-economic and environmental impacts to the designated operational entity (DOE) as a prerequisite for project validation (see Table 2-1). Host governments are expected to deny approval to projects that do not foster the country's sustainable development goals.

2.3. Carbon Sequestration Projects under the Broader Framework of 'Payments for Environmental Services' (PES)

From a broader perspective, LULUCF projects designed and implemented under the CDM are a special case within the more general framework of 'payments for environmental services' (PES). The central idea of such payments is that external beneficiaries of environmental services directly pay local landholders and land users for the adoption of practices that secure ecosystem conservation and/or restoration (WUNDER (2005)).

Besides the key role that forests play in the global carbon cycle, they provide many goods and services that society highly values. These include timber and non-wood products (e.g., fuel and fiber) as well as food and medicines; also biodiversity, soil and water conservation (e.g. protection against erosion) and recreational resources are important. PES schemes aim to seize the benefits of services such as regulation of hydrological flows, conservation of biodiversity, provision of scenic beauty and storage or sequestration of carbon (LANDELL-MILLS and PORRAS (2002)). The objective of such payments, ideally channeled to the managers of natural resources on the ground, is to reward them for service provision and to provide an incentive to do so in the future. Developing countries for example administer the majority of the world's tropical forests, which is notably an impressive carbon reservoir. By taking care of those environmental resources and managing them adequately, they provide a vital global public good by maintaining global environmental assets (STIGLITZ (2005)).

With increasing pressure on development and use of resources, developing countries can hardly be expected to provide these services for free. Thus, it is understandable that tropical countries in the theory of PES should be compensated for maintaining their tropical forests, thereby providing a very important global service (STRECK and SCHOLZ (2006)). Typically, land users receive no compensation for the service they generate for the general public, resulting in very little incentive to provide these services given prospective land uses which are economically more attractive (e.g., conversion of native forests to pasture or palm-oil plantation). This fact is one of the major reasons for human-induced deforestation (KAIMOWITZ and ANGELSEN (1998)). Governments in the past often have responded to the problem through public ownership of forests. However, raising awareness of the issue and the threats to service supply in the past decade have promoted ideas to develop new compensation systems like PES (PAGIOLA et al. (2002)). In addition, failures of former approaches based on command-and-control to address the aforementioned problems supported the introduction of more innovative and economically efficient systems.

A recent review of PES schemes identified 287 initiatives with the aim to reward forest environmental services worldwide (LANDELL-MILLS and PORRAS (2002)). PAGIOLA et al. (2005) argue that market-based mechanisms like PES are able to geographically concentrate efforts where either costs are lower (e.g., for conservation) or benefits are higher (e.g., for watershed protection). Although economically plausible, the actual targeting of PES programs (i.e., the site selection) remains a major issue to be addressed in the future (WUNDER (2007)).

A so called 'feedback loop' related to the structure of PES schemes (mainly based on experiences with watershed protection efforts in Latin America) has proven to be advantageous (ECHEVARRÍA et al. (2003)). This feedback loop ensures that disbursements to service providers are directly linked to the service users' payments. In this case the latter are eager to ensure that payments are spent effectively and if not, will most likely request the program design to be modified accordingly. While this is feasible in payment schemes based on watershed management activities, the situation is different for carbon storage or sequestration activities. Watershed management activities are characterized by the fact that both, service providers as well as service users regionally live in the same catchment area while in the case of carbon-related services the two parties involved are usually located much more disjunctive. Thus, given the missing spatial relatedness the described feedback mechanism will have to be replaced by appropriate institutional arrangements which can ensure that the incentives of service providers and service users remain aligned. However, PES schemes as well as the rules and regulations of the CDM described in the previous chapter are simply too recent to fully assess if they can actually meet this requirement (PAGIOLA et al. (2005)).

As outlined, PES schemes exist in four areas today. While those services are in principle all marketable the degree of actual feasibility varies from provision of clean water (regulation of hydrological flows), to carbon storage and sequestration, biodiversity conservation, and recreational aspects. Most literature exists with regard to water services followed by carbon-related services (LANDELL-MILLS and PORRAS (2002)).

In this context, it is noteworthy that the regulations of the UNFCCC clearly distinguish between services related to carbon conservation (i.e., protection of an existing carbon reservoir) and carbon sequestration (i.e., active storage of carbon in newly established trees). While the former are excluded from the project-based mechanisms and emissions trading up to this day, the latter are acknowledged in the form of afforestation and reforestation (A/R) projects under the CDM. The issue of carbon conservation is usually referred to as 'avoided deforestation' or, more recently, 'reduced emissions from deforestation and forest degradation' (REDD) in the context of the international climate regime (UNFCCC (2006a); THE WORLD BANK (2007)). In particular European non-governmental organizations (NGOs) have opposed the inclusion of avoided deforestation in any climate regime from early on in the international negotiations (FEARNSIDE (2001)). According to their views, climate change is a problem that should be addressed primarily through reducing the world's dependence on fossil fuels and cutting emissions at their sources. Hence, the inclusion of land-use related aspects in the climate debate, i.e. the inclusion of carbon 'sinks', would distract attention from that objective. Crediting countries for avoided deforestation projects would allow them to implement compensation measures that would let them off the hook in respect of more complex energy projects. Also, the more technical problems of carbon accounting and monitoring, the issue of 'leakage' (see the section on leakage above) and the possible non-permanence of carbon sinks would introduce intolerable insecurities into the system (RICHARDS and ANDERSSON (2001)). Many EU governments integrated these views into their official negotiation positions during the yearly COPs. Declaring to be motivated by environmental concerns, they adopted the partly simplistic views of critical NGOs, putatively because their campaigns carried a great deal of credibility within the public. Therefore at present, neither the UNFCCC nor the Kyoto Protocol include a satisfying mechanism to reduce the substantial climate-relevant emissions from deforestation, though they are responsible for about a quarter of global greenhouse gas emissions (DEFRIES et al. (2002); HOUGHTON (2003)).

Even though carbon sequestration services can be paid for under the CDM in form of A/R projects it is important to note that other project categories like sustainable agricultural practices (e.g., conservation farming) are in principle feasible and eligible from a PES perspective. However, in the "Marrakech Accords", the UNFCCC Parties for political reasons have limited the broader scope of LULUCF projects in developing countries to only A/R projects and further limited their use to meet only one percent of Annex I countries Kyoto obligations (UNFCCC (2001)). From an economic perspective it is therefore doubtful if the design of the CDM as it is set out today can meet its twin objective of carbon mitigation *and* sustainable development given its rather narrow scope on only A/R projects. Particularly developing countries highly rely on activities in the agricultural sector, which contributes substantially to the structure of their output.

It is widely acknowledged that growth in the agricultural sector is an engine for economic development (MELLOR (1966)). DELGADO et al. (1998) supported this finding in particular for developing countries through empirical research in Sub-Saharan Africa. Given the fact that developing countries are following a pattern of growth with agriculture paving the way for other economic activities, additional revenue streams through LULUCF projects in the agricultural sector are of particular interest. NILES et al. (2002) point out that the potential income in developing countries from carbon crediting for changes in use and management of agricultural and forest lands could be significant. Even under current uncertainties with regard to future carbon prices, carbon finance is considered as non-deniable contribution to the income of rural communities. It is, however, questionable if the CDM will be able to meet its striving twin-objective with the current restrictions applying.

PES schemes, in contrary, never aimed at such ambitious, twofold goal. They were primarily designed as a mechanism to improve the efficiency of natural resource management (see WUNDER (2005); ECHEVARRÍA (2002); ROJAS and AYLWARD (2002)). Moreover, fundamental doubts exist about the extent to which it makes economic sense to link the conservation and development agendas under one single mechanism (ADAMS et al. (2004)). While environmental service provision is more and more discussed as potential new means for the simultaneous pursuit of rural development and natural resource conservation particularly for landowners in the low-income tropics, some preliminary evidence suggests that participation of, and benefits to, small-scale land users can be highly unequal. TSCHAKERT (2007) for example argues that the synergistic goals of poverty reduction and resource conservation through these emerging market mechanisms might be overly ambitious.

On the other hand, an emerging body of literature is pointing to the substantial synergies that might exist between PES, rural development and poverty reduction (GRIEG-GRAN et al. (2005); PAGIOLA et al. (2005)). These potential synergies exist because a large share of people living on 'fragile' and/or 'vulnerable' lands also suffer from extreme poverty (THE WORLD BANK (2003); SUNDERLIN et al. (2005)). Those fragile lands, including frequently sloppy terrains in the upper reaches of catchment areas, poorer arid zones, and forest ecosystems are typical target areas for PES (see for example XU et al. (2004)). The intended recipients of PES schemes do mostly live below the poverty line which brings about the environment-poverty link. At all events and particularly given the different views in the scientific literature, the design of PES schemes necessarily has to deal with the question of poverty alleviation and whether markets offer opportunities or obstacles for marginalized groups to participate.

2.4. Carbon Mitigation and Potential Developmental Benefits of LULUCF Projects – Some Prospects and Potential Pitfalls

Forestry projects are a relatively low cost and rapidly implementable strategy for at least temporary climate change mitigation with a long-term sustainable development component that can considerably contribute to poverty alleviation (LIPPER and CAVATASSI (2003)). This is due to the fact that approx. 70% of the world's poor are located in rural areas, with agriculture and forestry being a major source of their subsistence (FAO (2000)). Thus, forest-related projects are a direct and feasible investment at the local level that does not require a complex capacity building up front like other CDM project categories (e.g., in the energy sector). However, an emerging body of literature especially in the field of PES raises concerns about unintended impacts of carbon service payments with regard to poverty alleviation (LANDELL-MILLS and PORRAS (2002); TSCHAKERT (2007)). This section assesses some prospects and pitfalls of the linkage between environment and development agendas in PES schemes and the CDM and discusses the most important issues to look at.

Under the CDM framework, the COP decided that it should be the CDM host Party's (i.e., the developing country's) sole right and prerogative to decide whether a CDM project activity contributes to 'sustainable development' or not. It is axiomatic for the CDM that the host country defines and interprets the general term 'sustainable development' under its specific circumstances and priorities (UNFCCC (2005b)). Therefore, host governments are expected to deny approval to projects that do not promote their country's sustainable development goals. For example, LANDELL-MILLS (2002) argue that governments charged with ensuring equity as well as economic efficiency and environmental sustainability need to generally pose the critical question how the rise of markets for environmental services impacts on the poor. The same is true for the CDM: how does the goal of carbon mitigation influence project-related socioeconomic parameters, and ultimately poverty? Asking this question is not only of concern for ethical reasons, but marginalization of the poor from the market development process could have critical implications for the sustainability of market systems, e.g., undermine their effectiveness in achieving environmental objectives (LANDELL-MILLS (2002)).

In order to assess the overall impact of markets for environmental services, several aspects have to be considered individually. First and foremost the issue of land ownership and land tenure needs to be looked at in more detail. Secondly, transaction costs of project implementation as well as the equal distribution of costs and benefits need to be addressed. Furthermore, conflicting target approaches between natural resource management and socioeconomic characteristics of the service providers (e.g., poverty) have to be considered.

The fact that in principle only land-owning or land-controlling farmers can participate in PES schemes as well as in LULUCF CDM is on the one hand a rather general but on the other a fairly important finding (ANGELSEN and WUNDER (2003)). PES payments are in essence payments to land users since the services paid for are often the result of a particular kind of land use. This directly points to the critical issue of land ownership and land tenure which implies that landless people who are often the poorest of the poor are excluded from the program by its entry regulations (GRIEG-GRAN et al. (2005)). Furthermore, farmers with only some hectares cannot set aside land for the PES or CDM-related activity if food production for subsistence of the family is their major occupation. In a review of eight Latin American PES projects GRIEG-GRAN et al. (2005) found that at least two projects had eligibility criteria requiring a certain size of land holdings. In other situations an unambiguous land title was an eligibility prerequisite. Thus, people with only informal tenure were excluded from the project activity. Even though the lack of titles should not be equated with tenure insecurity (e.g., titles may not be necessary as long as tenure is secure, see PAGIOLA et al. (2005)), it is important to note that insecure tenure is often associated with lacking access control which in turn is an unreliable precondition for the service provision in general. If access cannot be controlled, specific land management services cannot be guaranteed which makes the service provider unreliable.

In the presented Tanzanian case study we will discuss in the subsequent chapter that livestock encroachment and grazing turned out to be a problem for the trees established for carbon sequestration purposes. Especially in situations where PES programs require long-term investment decisions, the issue of tenure security is of particular importance. This is the case for reforestation project activities with its long-term gestation periods for the investment. Additionally, possible access restrictions imposed by the project design might worsen the livelihoods of herders and other landless poor who depend on gathering non-timber forest products. As a result, access restrictions that might be advantageous with regard to program implementation might actually negatively impact on other social groups.

From an economic perspective, the value of currently marginal, fragile, often rather unfruitful sites is increased through a PES program or a CDM project. This may lead to intensified competition for the formerly valueless land. The associated problem is that the more influential groups within a rural society will take control of the land if there is no pre-defined tenure security. Social tensions between participants and non-participants of the PES or CDM project activity will most likely occur as an unintended result.

Another critical issue to look at are transaction costs. Transaction costs reflect the time, effort and other resources needed to search, initiate, negotiate and complete a PES or carbon finance deal under the CDM, i.e.,

a) the costs for market participants to find one another,
b) costs of demonstrating additionality and assessing leakage,
c) costs of monitoring and
d) costs of enforcing contracts etc. (GUTIERREZ et al. (2006)).

Transaction costs lower gains from doing business for both, carbon buyers, as well as carbon sellers. In this regard, two forms of transaction costs have to be distinguished: (i) the costs for the project developer of contracting with each participant and (ii) costs imposed on participants themselves.

Several studies point to the plausible fact that transaction costs per unit of emission reduction are higher in projects involving local communities (SMITH and SCHERR (2003); BASS et al. (2000); HARDNER et al. (2000)). Not only the number of potential participants is challenging in such projects but also their remoteness, i.e., their scattered settlement patterns. In addition, high transaction costs arise from different land uses and management practices on a small scale and often a higher degree of uncertainty in the rights of small farmers of their land-based property. There are clear economies of scale effects favoring the purchase of certain services from a very limited number of landowners (e.g., carbon sequestered from only one privately-owned plantation) rather than negotiating contracts with hundreds of smallholders.

To overcome this problem, GRIEG-GRAN et al. (2005) suggest to develop contract designs that allow for collective agreements or group contracts. This implies that land is operated collectively or at least on a basis of very close co-operation amongst participants. In such a situation, the decision to participate in the project from the seller's side must then, in principle, be made and implemented collectively. Whether the resulting costs and benefits of the project are distributed equally within the group then becomes a major issue. Clear competencies and rules of leadership within the group need to be developed and put down (i.e., adequate institutions need to be established) in order to ensure that payments trickle down to the actual service providers on the ground, namely the farmers. Even within the group underprivileged farmers are frequently not positioned to lobby for their interests, which bears the risk that better endowed farmers become more assertive and dominate the group's interests. In the case study assessed in more detail in the subsequent chapter, transaction costs are reduced through the creation of so called "Small Groups" of farmers. As will be described, these groups consist of five to twelve farmers, which jointly sign a contract for the provision of carbon sequestration services (see further down). As in other cases, transaction costs in the Tanzanian project are reduced through the activities of an intermediary. Most frequently such an intermediary is an NGO or NGO-like institution (CACHO et al. (2003)).

Another issue to look at is the targeting of the PES scheme or CDM project activity. PAGIOLA et al. (2005) rightly state that the geographic, natural resource-based targeting employed in a PES program cannot prioritize poverty. This is certainly true for PES schemes where certain areas provide specific services, i.e. regulation of hydrological flows and provision of clean water. However, with regard to carbon sequestration, be it under a PES scheme or a CDM project, the case is somewhat different. The upstream downstream problematic is not given in carbon sequestration projects. These projects are not spatially specific, in principle. This means that reforestation projects can theoretically be implemented 'anywhere' on the globe. Thus, on the one hand, carbon-related projects could be much more poverty-targeted. On the other hand, it is important to be aware of the fact that smallholders have no specific comparative advantage with regard to their location to differentiate their services from service providers of larger scale. Site selection in carbon markets might eventually become very competitive which counteracts the poverty orientation in the end.

Besides those general considerations with regard to the program design, PAGIOLA et al. (2005) outline three key questions to address in order to better understand the potential developmental benefits of environmental service payments with regard to poverty. These are:

1. Who are the actual as well as potential participants of the scheme and how many of them are poor?

2. What are obstacles for the participation of poor farmers in PES, or framed differently, which factors influence their participation decision?
3. And, what are the actual impacts of the payment scheme on participants?

The first question concerns the identity of land users actually involved or targeted by the program. Substantial variability in actual socioeconomic levels can exist among participants as well as non-participants who might potentially become program participants in the future as the program affects their residence area.

The second question, addressing factors that influence the participation decision, is the main focus of the study at hand. As PAGIOLA et al. (2005) point out little knowledge exists up to this date with regard to the program participants' actual participation decisions. Most PES as well as carbon finance projects are just too recent for an assessment of participation decisions. However, the study at hand gains insights from literature that examines factors determining participation in other rural programs like land conservation. Findings of these studies have to be tested and evaluated analogously under PES and carbon finance project situations.

In order to address the third question with regard to the project impacts on participants it is useful to differentiate between income-related impacts and non-income ones. In general, solely the existence of a payment to participants is not a good measure for any financial benefit the project participants might receive. Instead, the measure to be used preferably is the amount of payment net of the opportunity cost of adopting the PES or carbon finance induced land use (FERRARO and SIMPSON (2005)). In any case, the amount of payment has to outweigh the opportunity costs of the next best land-use alternative. At this stage, different opportunity costs of different land users come into play, which indicates that 'flat payments' are in general questionable from an efficiency point of view (PAGIOLA et al. (2005)). Nonetheless, under a flat payment, poorer farmers who have the lowest land opportunity cost might incur the greatest producer's surplus. Opportunity costs are usually difficult to assess while it is easier to survey the net payments. For that reason, PAGIOLA et al. (2005) conclude that given participation of the farmers in the project or program is voluntary it will make them better off ('prima facie assumption').[11] Analytically, this might work as a starting point since data on the possible increase of household income amongst participants caused by income from PES are rather sparse.

Turning to the analysis of non-income program impacts LANDELL-MILLS (2002) suggests to evaluate how markets impact on the broader asset base on which poor farmers depend. A number of such asset-based approaches have

11 However, the term 'voluntary participation' has to be questioned with caution. Sometimes participation is only voluntary in parts, see for example the description of XU et al. (2004) which reveals that membership in the Chinese 'Grain for Green Program' was officially 'voluntary' but actually some farmers were not even asked by the administration before they were enrolled in the program.

been developed under the general topic of 'Sustainable Livelihood Approaches'.[12] Table 2-2 presents six different asset classes to be evaluated in order to assess rather non-income related program impacts. Except for the category of 'financial assets', all other categories (i.e., natural assets, social assets, human assets, physical assets, and political assets) aim to assess a broader, livelihood-related picture.

Table 2-2 Potential Impacts of PES / Carbon Finance Payments on Assets of Poor Households

Opportunities	Risks
Natural Assets	
Increased value of marginal lands due to new market opportunities. Positive spill-over effects for timber and non-timber forest products (NTFPs) where sustainable forest exploitation is permitted. Positive impacts on other related natural assets (i) soil fertility and agricultural yields (ii) water flows and quality (iii) air quality.	New markets inaccessible by poorer farmers due to uncertain property rights, insufficient marketing skills and education, inadequate information, lack of contracts, inadequate communication structure, inappropriate commodity design, e.g., long-term contracts unsuitable for livelihood strategies that require short-term flexibility. Opportunity costs where delivery of the environmental service prohibits / hinders production of timber and / or NTFPs.
Social Assets	
Market may foster the formalization of resource tenure and clarification of property rights over environmental services.	Increased competition for control over land, social tensions.
Increased organizational and management capacity of community-based organizations through collaboration in delivery of environmental service.	Splitting up of communities in households gaining and households loosing due to the new market situation. Potential threat to local culture as environmental services become commercial assets.
Human Assets	
Improved education and skill base relating to forest management for environmental services, cooperation project management, marketing, negotiation, enterprise development, etc.	Poorer households excluded, can only work in subordinate jobs, necessary skills are brought in from outside.

12 A 'Sustainable Livelihoods Approach' (SLA) is a way to improve the general understanding of the livelihoods on which poor people depend. It draws on the main factors that affect poor people's livelihoods and the typical relationships between these factors. SLAs can be used in planning new development activities and in assessing the contribution of existing activities to sustaining livelihoods (CHAMBERS and CONWAY (1991)). A concise and useful overview of several different 'SLAs' is provided by CARNEY (2002).

Improved health due to more varied diets, improved water supply (both quantity and quality), improved air quality and microclimate, increased disposable income for medical treatment.	Reduced health due to lost access to NTFPs and associated nutrition and reduced disposable income.
Physical Assets	
Investment in improved communication links to remote forest areas.	Improved communication channels only available to certain market participants leading to few spin-offs for the rest of the community, inequality increases.
Financial Assets	
Diversified income base, regular income with prospect to be paid over many years.	Market regulations excluding poorer farmers lead to negative repercussions on their financial assets.
Political Assets	
Improved community-based organization provides basis for gaining political representation and voice.	If markets lead to marginalization of poorer farmers, these have even fewer channels to influence political decisions.

Source: LANDELL-MILLS (2002) - modified

Critics of this approach have rightly pointed to the fact that the different asset categories in Table 2-2 are partly overlapping. In addition, ANGELSEN and WUNDER (2003) note that the principal goal of the approach, namely to measure welfare gains, should rather focus on returns to those assets and that ambiguous situations with partly improving, partly deteriorating assets can not sufficiently be captured. Taking this valuable analysis into account, the framework provided by Table 2-2 is taken here as a starting point for project assessment with regard to potential development benefits and their possible trade-offs. Some of the aspects summarized in Table 2-2 have been discussed above in more detail; others will be revisited and commented in depth in chapter 6.1 where asset-based project impacts of the concrete case study are reviewed.

Besides the 'service providers', which are the farming households actually planting trees for the purpose of carbon sequestration (i.e., the participants of the program), other groups can be affected as well. Amongst them are poor service users (in case of watershed protection), poor landless laborers not engaged in farming activities predominantly found in the project area, and poor consumers, or other groups affected by the project. With regard to the service users GRIEG-GRAN et al. (2005) point to the fact that the climate mitigation activity of project participants (i.e., the tree planting) might actually facilitate adaptation to climate change of other poor farm households (i.e., the service users). The latter are benefiting from the positive impact on global warming through carbon sequestration on a macro scale.

On a more local scale positively affected micro climates (e.g., increased precipitation, wind breaks, erosion control) might be of benefit for 'external' service users as well (LOS et al. (2006)). Poor landless laborers might benefit from the project if the scheme increases employment options through a labor-demanding activity. In the case of tree planting a positive employment effect can be expected. But if in contrary activity levels are reduced through the project, employment benefits are reduced as well. For poor consumers and other groups GRIEG-GRAN et al. (2005) show that if a relatively large area is set aside for the program activity, the regional production of staple crops could be affected, which in turn could drive up prices to the disadvantage of poor urban consumers of the same crop. PAGIOLA et al. (2005) however point out that such effects appear quite unlikely as areas enrolled in PES programs are relatively small, so far. The same is true for the carbon finance projects actually registered by the Executive Board of the UNFCCC.[13] Though, other downstream effects of changes in production structures of other goods that may affect the poor have to be taken into account as well.

13 See page "Methodologies for afforestation and reforestation CDM project activities", http://cdm.unfccc.int/methodologies/ARmethodologies, accessed 30.08.2006.

3. Case study: The International Small Group & Tree Planting Program (TIST)

The following chapter gives an overview of the case study setting and the field research conducted for data collection. At first, the set up and purpose of TIST, the International Small Group and Tree Planting Program is described in more detail. Subsequently, the agricultural household survey for data collection, which was conducted from March through June 2005 in Tanzania, is portrayed. Special emphasis is put on the sampling procedure and the survey set up in order to ensure representativeness of the study findings. Finally, descriptive results of the case study are presented with a focus on interesting details as related to the data structure and hypotheses.

3.1. TIST: Program Design and Purpose

Data for the case study was collected in Tanzania in *"The International Small Group and Tree Planting Program"* (in short: TIST). The TIST program started off in the town of Mpwapwa in the central Dodoma region of Tanzania (see map). Initially in 1999, the program was an idea spread by the local Anglican Church. In the early stages of TIST the church motivated a process that encouraged rural farmers to identify their primary needs and to develop ideas how to overcome major shortcomings in the livelihoods of their families. The process resulted in the formation of 40 so called 'Small Groups' of farmers, each consisting of five to twelve members. The basic concept was to establish a kind of safety net through the formation of the groups to aid one another, to prevent the risk of famine and thus reduce poverty. A joint effort to reforest degraded lands was part of the initial concept as well. As most of the district forests of the town of Mpwapwa were depleted during the second half of the last century farmers were convinced of the reforestation effort. In the following years, project development was backed and partly subsidized by 'Clean Air Action Corporation' (CAAC), an American organization with a background in the NO_x, SO_x and greenhouse gas trade in the US.[14]

According to CAAC the company would not have invested in the TIST program but for the expectation of an emerging carbon market, and the prospect of sale of GHG credits stemming from the project (BIOCF (2006)). Although the international carbon market in the forestry sector has been slower to mature than originally anticipated by the project sponsors, the program expanded rapidly from 2000 to 2004. At some point in time the program even comprised four different project areas in four Tanzanian regions with more than 10,000 subsistence farmers participating.

---

14 For more details about Clean Air Action Corporation see http://www.cleanairaction.com. Furthermore, USAID and the Dow Chemical Foundation supported the project development.

However, the apparent success and extremely rapid expansion of the program also caused problems. Low survival rates of planted trees and structural problems of the program design resulted in delayed delivery of carbon finance payments to participants. Normally, Small Groups are supposed to receive quarterly cash payments from CAAC according to the number of living trees maintained by the Group, notably as an incentive to tend the planted trees. One major reason for the pilot phase problems was the sheer size of the program due to its rapid expansion, its geographical scatteredness, and the remoteness of participating farm households (FRANK (2005)).

A subsequent shrinking process of the program followed. Such a 'bushfire' phenomenon coupled with a following shrinking process can be observed in other PES programs as well.[15] Today, TIST is present in two different regions of Tanzania with approx. 240 Small Groups participating. Now, that the problems of the pilot phase have been mostly solved, it has also spread successfully to Kenya, Uganda and even to India.[16]

Finally, the project applied to become one of the World Bank's so called 'BioCarbon Fund' projects. The BioCarbon Fund bundles mainly CDM forestry projects in developing countries. Its purpose is a) to act as a pooling entity to lower transaction costs of carbon transfers, b) disseminate information to both sides, project developers and buyers of carbon certificates and c) to re-distribute - on a discretionary basis - acquired credits to industrialized countries that have taken on emission reduction commitments.

The TIST program goals are manifold but focus on the combination of sustainable rural development and carbon sequestration. On the one hand, farmers involved in the project plant indigenous trees, which provide them with fuel wood, lumber, fruits and shade. The tree planting stabilizes erosion, improves the soil and ultimately the agricultural yields. In addition, the provision of shade and windbreak by the trees improves the microclimate in the villages where the tree groves are. On the other hand, the tree planting activity generates income through GHG sales.

The planted trees are capturing carbon dioxide from the atmosphere through photosynthesis in order to grow. They store carbon in their woody biomass and in the soil. The amount of carbon stored in the trees once quantified can be certified and subsequently sold on the international market. Thus, the trees create a 'virtual cash crop' in the form of GHG certificates that can be sold to buyers in industrialized countries (Annex I countries). The sale of these carbon certificates can thus constitute a long-term source of regular income for farmers and is supposed to ultimately help to improve livelihoods.

---

15 For a description of such a rapid program expansion, its associated problems and the later shrinking in the Chinese "Grain for Green" program see XU et al. (2004).

16 The TIST program is featured at www.tist.org. Select "Project Areas" and "Tanzania" to view the current number of Small Groups, trees and seedlings, including the precise latitudes and longitudes of smallholder groves.

Besides the external verification of the amount of carbon stored, every five years due to the international regulations outlined above (see chapter 2.2), carbon quantification within TIST is done by so called 'quantifiers'. They are trained within the program. These quantifiers are farmers or non-farm workers from the local area that make visits to the Small Groups. The groups' trees are surveyed and assessed by counting number, height, and tree species. In the field, the actual geographic delineation is logged by GPS location into a PDA-based database. Thus, data can be stored and managed locally and also exchanged internationally.

The quantification and reporting infrastructure of TIST was recently expanded by the use of hand-held computers with integrated digital cameras. Both, GPS and hand-held computers facilitate transparent and timely reports. In addition, CAAC staff provides regular computer training for TIST staff in Tanzania in order to manage the data and coaches local TIST staff in business skills for the administration of the project. At the time of the field research, the project employed 16 full-time Tanzanian nationals that have been recruited from the TIST area.

Besides the tree planting activity, TIST encourages farmers to shift at least part of their production system to so called 'conservation farming' techniques. Those techniques imply minimal soil disturbance through reduced or zero-tillage systems, permanent soil cover and the planned rotation of crops (IIRR and ACTN (2005)). In addition, the program provides training on the development of tree nurseries and other topics like the use of fuel-efficient stoves, or nutritional education.[17] Furthermore, the program supports preventative health education including HIV/AIDS, and malaria. This training takes place at regular intervals for several Small Groups working in the same area. Thus, the implementation of the two main components of the program, carbon sequestration and sustainable development, takes place on the Small Group level. It is the organizational unit of the Small Group that is supposed to provide the mutual support and accountability that assures compliance with program requirements. Farmers commit themselves for at least five years to participate in a TIST Small Group by signing a covenant that includes the following commitments:

- Set aside 10% of its agricultural produce as an emergency fund for other group members

17 For the development of nurseries Small Group members collect seeds from local trees. No invasive species are introduced by the project, and the only 'non-native' species planted were already long introduced in the project area. The planting is usually done at the beginning of the rainy season. Prior to the rainy season, holes are dug for the trees using simple hand hoes. At the beginning of the rainy season, the seedlings are then transferred from the nursery to the holes. Usually the trees receive enough water in the rainy season to grow to a height enabling them to survive through the following dry season. This method improves the survival rate of the trees and reduces the amounts of water needed for the planting.

- Follow and help develop "best practices" in the planning of trees and the development of Small Groups
- Use "best practices" developed to plant, care for, and establish an agreed amount of trees per year. Each group is responsible for protection of all other TIST group trees that have been or will be planted, wherever they are planted
- Enter into an agreements to transfer the GHG credits created from the TIST trees
- Meet regularly in the Small Group (once per week), and to rotate the leadership of the Small Group
- Report group activities to TIST on a regular basis and to cooperate with verification efforts necessary to establish GHG credit creation

A typical Small Group planting and maintaining 2,000 trees may at present earn around US$40 per year through the quarterly flat rate carbon finance payments of CAAC (BIOCF (2006)).[18] Given this current order of magnitude of carbon finance payments, increased crop yields due to conservation farming techniques are supposed to outweigh the carbon finance component of the program.[19] The program envisages to channel carbon sale proceeds to the local villagers by a network of rural banks throughout Tanzania. However, given the relatively high immobility of farmers due to the lack of transportation in remote rural areas, this remains a major problem to be addressed in the future, as we will discuss further down.[20]

With regard to the amount of carbon sequestered in the project, first estimates of carbon storage were around 3.5 Mt CO_{2e} by the year 2012 (the end of the first commitment period of the Kyoto Protocol) and about 4.9 Mt CO_{2e} by 2017 (end of assumed second Kyoto commitment period) (BIOCF (2004)). The entire project life is projected to exceed 75 years, but this would be clearly dependent on the existence of a carbon market in the agricultural sector in the future.

18 Assumptions: average Small Group: 10 members, 2000 trees on 4 acres, application of conservation farming to one acre per member.

19 With the carbon market maturing, in the future carbon finance payments to farmers should be linked to actual market prices, which will most likely increase the amount of payment per year.

20 During field research it was discovered that the huge majority of farmers very seldom leaves their ancestral village. This is of special interest given the fact that the opening of a bank account - a requirement for groups to participate in the program – can very often only be done in larger rural villages, which are better connected. Farmers moaning about delayed carbon finance payments were asked for their visiting quantity of the next larger village. Interestingly, in 90% of the cases farmers visit the larger village only once per year to see relatives during Christmas time. However, none of the banks operating in the research area was open during Christmas.

3.2. The Agricultural Household Survey – Sampling Procedures and Data Collection

Data for the empirical analysis was collected in mid 2005 by conducting an agricultural household survey. The survey was planned and designed as 'multi-topic' household survey by collecting data on various aspects hypothesized to influence the decision making process of the farm household (e.g., labor force, household asset and resource endowment, family structure etc.). The design of the questionnaire drew on the experience of the World Bank's Living Standards Measurement Study (LSMS) program (GROSH and GLEWWE (2000)).[21] Like in a LSMS, the overall objective of the questionnaire was to measure the determinants of living standards in the survey area of rural Tanzania with a special focus on the living standards of the poor who are actual or potential TIST participants. To achieve this goal the survey gathered information on several aspects determining living standards, like the economic and social environment in which household members live. On the choices that households make several variables were retrieved. Similar to an LSMS two different questionnaires to collect information about various aspects of household welfare and behavior were elaborated. A household questionnaire and a community questionnaire included a substantial part on prices of commonly purchased items on local markets.[22] In addition, the community questionnaire gathered information on the resident setting common to all households living in the same community. This includes information on a) weather conditions, b) the existence of local credit institutions, c) the most important employment opportunities, and d) local agricultural conditions and practices. Still, the main part of the survey was the household questionnaire which generated detailed information on

- the household composition (household roster),
- education, and employment,
- consumption as a monetary indicator of household welfare,
- housing conditions,
- agricultural production,
- household enterprises,
- miscellaneous income sources,
- availability of credit etc.

21 The Living Standards Measurement Study (LSMS) program was established in 1980 by the World Bank in order to explore new ways of increasing the accuracy, timeliness, and policy relevance of household survey data collected in developing countries. Amongst several efforts by international institutions to expand the pool of data on poverty and living standards in developing countries, the LSMS program is one of most methodologically sound approaches.

22 While a LSMS is often based on a separate price questionnaire these aspects were incorporated into the community questionnaire prior to the field trip in order to speed up and facilitate fieldwork.

Some of the information was collected at the household level (e.g., consumption, housing quality etc.) while other information was surveyed in more detail at the individual level (e.g., employment, education) or even on the plot level (e.g., soil quality, slope, tenure rights and access restrictions, agricultural production etc.). The questionnaire comprised 14 different parts (modules), one of them being a special module on TIST participation (see Appendix 4). The modules were synchronized with each other in order to detect gaps and overlaps. All potential answers were coded and the questionnaire wording was harmonized prior to fieldwork.

Subsequently, both questionnaires (household and community) were translated from English to Swahili and field-tested. For this purpose, five master students from Sokoine University of Agriculture in Morogoro, Tanzania who had gained previous experience in the design and implementation of agricultural household surveys were specifically trained as interviewers. Survey specific training elements comprised for example carbon finance projects under the Kyoto Protocol in developing countries. In addition, the interviewers familiarized themselves with the design and purpose of the TIST program.

During fieldwork, interviewers were supervised every day and about one third of all interviews was directly observed. Questionnaires were immediately reviewed after the interview in order to correct any discrepancies in the documents right on the spot. In addition, the same interviewer team did data entry and editing as soon as interviews were finalized. An interface-based MS ACCES® database was prepared prior to the field trip so that data could be easily entered via a mask carrying out a large number of quality checks to detect responses that were out of range or inconsistent.

As mentioned before, information was gathered on more detailed levels than the household as far as possible (e.g., employment of the different household members, education of household members etc.). By avoiding asking one single person all the questions about the entire household but rather conducting a series of 'mini-interviews' with each adult responding for himself or herself, errors caused by respondent's fatigue or by the use of proxy respondents were minimized. Special focus was, however, on the head of the household during all interviews, assuming that decisions for the whole household depend to a very large extend upon his or her capacities.[23]

All the data collected refer to farm system, household characteristics, and production decisions occurring in the calendar year 2004. The 'mini-interview' design was also very helpful in mitigating the problem of recall bias inherent in such a retrospective research design, as household members were able to remind each other and help each other out with information during the interviews. Even though some authors suggest recall periods of only a few months given the sea-

23 Extending the basic household model to explore the importance of bargaining power of various household members in decision making is at the frontier of current research (see BEHRMANN and RAYLYNN (2000)).

sonal economic environment of peasant farmers, which tends to influence the reliability of responses (see CAVENDISH (2002)), others suggest that recall periods of one year guarantee accuracy of the responses which was also found to be a good compromise for the actual field survey (GROOTAERT (1982)).

3.2.1. Sample size

The sample target envisaged gathering at least 200 utilizable data sets of farm households living in 20 different villages in the eastern and central zone of mainland Tanzania. The aim was to split the dataset evenly between households participating in TIST and non-TIST participants. The actually required sample size was established on the following assumptions. The sample size in general depends upon three factors.

- First, the level of precision required in the estimates. The more precise the estimates are required to be the larger the sample size must be.
- Secondly, the inherent level of variability of the variables to be estimated. If the characteristic to be estimated is itself naturally more variable in the population, a larger sample size is required to attain a given level of precision.
- And finally, the sample design to be used also determines the sample size. A stratified sampling design generally reduces the required sample size for a given precision. In empirical research however, especially in a developing country environment like in Tanzania, also more pragmatic deliberations with regard to survey costs, timing and transportation play a role in identifying the final sample size of a field study.

For the planned survey, at first, the acceptable level of precision was specified by setting an acceptable margin of error (i.e., the estimate should be correct within +/- 5%) and the related confidence level (most common confidence level: 95%).[24] Subsequently, all other factors influencing sample size as outlined above have to be taken into account, most importantly the variation in the population was reconsidered. The variation in the population, c, measures the inherent variability within the population of the variable being estimated. Normally this value is not known and the survey is actually seeking to find out the variation in the population. However, a rough idea can be obtained from results of surveys of similar variables in similar rural areas. A typical value for an agronomic value like for example the yield of major crop grown by peasant farmers under rain-fed conditions is in the range of 40-80% variation (POATE and DAPLYN (1993)). To estimate the mean of a variable, with a variation c (%), confidence level z, and precision x (%), the required sample size, n, is given by:

24 The chosen confidence level is expressed as a value of the z statistic, from the normal distribution. Value of z is 1.96 at 95% confidence level and 1.64 at 90% confidence level.

$$n = \left[\frac{zc}{x} \right]^2 \tag{3.1}$$

The reformulation of (3.1) to assess the maximum variation allowed in the variables collected in the survey leads to:

$$c = \frac{\sqrt{n} \cdot x}{z} \tag{3.2}$$

For the envisaged survey with at least 200 farm household data sets, a precision level of 5% and 95% confidence, equation (3.2) would mean an approx. 74% allowed variation of the surveyed variables. This variation was considered to be a reasonable variation estimate to work with (POATE and DAPLYN (1993)). In addition, the sample size had to match the requirements of the structural equation model as described in chapter 5.

3.2.2. *Administrative and geographical characteristics of the survey area*

The following section gives a brief overview of the administrative and geographical circumstances of the survey area.

The whole of Tanzania is divided into 26 administrative regions. These regions are further subdivided into 130 districts, which are further segregated into wards and finally villages. The area of the survey covered two Tanzanian regions, namely the Dodoma and the Morogoro region and three districts within those regions, which are Mpwapwa, Kongwa, and Kilosa. These two regions encompass approx. 113,000 square kilometers[25] and cover at least two different agroecological zones within the semi-arid lands (agroecological zone III) and the highlands (agroecological zone V) (TANZANIA VICE PRESIDENCE'S OFFICE (1999)).[26] The semi-arid lands are characterized by an elevation of approx. 600m to 800m above sea level and are located more in the northern and middle part of the survey area. The zone has a savanna type of climate, with low and erratic rainfalls in the range of 400 to max. 800 mm per year. It is characterized by a long dry season lasting between approx. late April and early December, and a short unimodal rainy season occurring during the remaining months. In the long dry season, persistent desiccating winds and low humidity contribute to high evapo-transpiration and topsoil erosion. The southern part of the survey area extends into the Rubeho Mountains, which reach an altitude of up to 2,200m above sea level (peak of Mang'alisa). Precipitation is higher in the highlands and

25 This is approx. equivalent to the size of one third of Germany.

26 The combined influences of altitude, latitude, rainfall and soil determine the agroecological zones of Tanzania. While PRATT et al. (1966) classify Tanzania into five agroecological zones, the seven-zone classification of the Tanzanian 'Proposed National Action Programme to Combat Desertification' (TANZANIA VICE PRESIDENCE'S OFFICE (1999)) is used here.

mountains (800–1,000 mm per year) with local rainfalls of up to 1,400 mm per year (TANZANIA VICE PRESIDENCE'S OFFICE (1999)). Fig. 3-1 portrays the survey area and the 20 selected villages, which are marked by light-gray rectangles.

Fig. 3-1 Map of the Survey Area in Central Tanzania (top of map is geographic North)

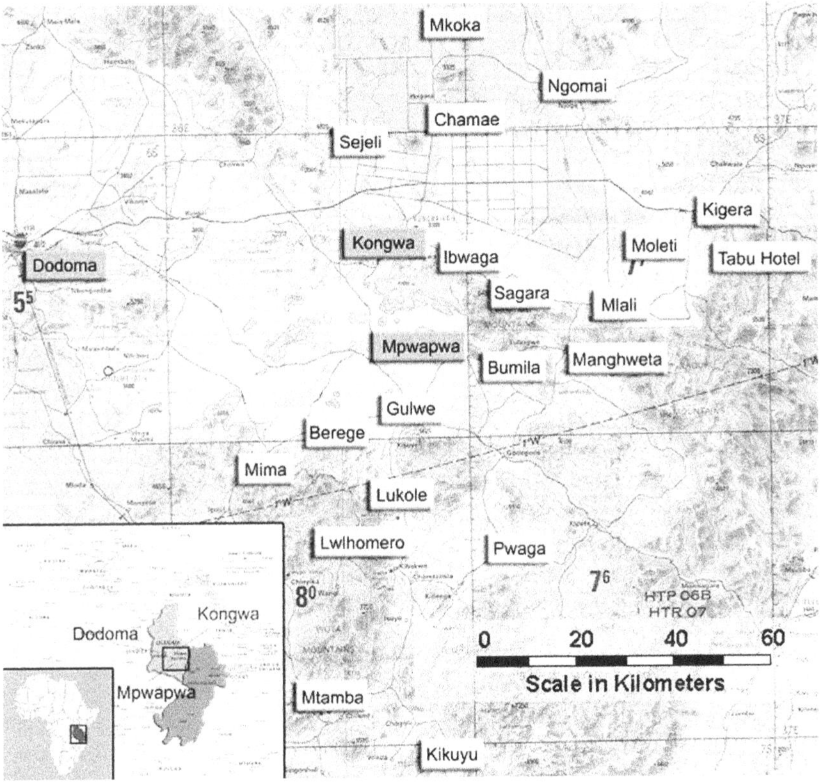

The villages Bumila, Gulwe, Berege and Mima in the middle part of the research area are connected to a non tarred rural road going from the district capital of Mpwapwa (dark-gray rectangle) to the Rubeho / Wota Plateau and than further down south to the regional capital of Iringa (not on the map). The other five villages in the south of the survey area (Lukole, Lwilomero, Pwaga, Mtamba and especially Kikuyu) extend into the Rubeho mountain range. Conditions of the rural roads worsen in the mountains.[27] The tarred, principal highway

27 Only approx. 4% of all Tanzanian roads are tarred or paved (THE WORLD BANK (2003)).

connecting Dar es Salaam and the administrative capital of Tanzania, Dodoma, transects the survey area in the northern part. The villages located most closely to the highway are Kigera, Tabu Hotel, Moleti and Sejeli. Chamae and Ngomai both are located approx. 20 to 30 km north of the tarred principal highway. And Mkoka a little bit further in the North.

3.2.3. Village selection and sampling of households

Our selection of the 20 villages on the regional level to conduct the household interviews was based on probability proportional to size sampling of the most recent National Agriculture Sample Census (NASC) of Tanzania (TANZANIA NATIONAL BUREAU OF STATISTICS (2006)). This information was merged with a regional database of CAAC indicating participation of households in the TIST program on the village level. The same methodological approach, primarily to merge two data bases and select villages based on superjacent information, was used by ISHAM (2002) in his adoption study on fertilizer use in rural Tanzania. The combination of the two data sets finally provided the 20 sample villages. On the village level the NASC provided a list of all farm households. Respondents were drawn from this list by 'systematic random sampling' (KALIBA et al. (1998)). Given the fact that villages in the NASC were selected with probability proportional to size sampling, the number of sample households per village in the second stage of sampling was kept constant (see POATE and DAPLYN (1993)). The population size (sampling frame) was divided by the required sample size per village (at first five farmers) so that every k^{th} household was drawn to be included in the sample. In cases where the selected farmer was not available or absent, the next household on the village household list was interviewed. Depending on the participatory status of the chosen farmer (TIST member or non-TIST farmer), a complementary farmer was drawn through 'randomized pairing' with the former subject based on decision rules regarding ordinal direction (e.g., next non-participating farmer to the south of the participating farmer, see ZBINDEN and LEE (2005)). Thus, the final sample size per village was ten farmers with five farmers participating in the TIST program and five non-participants.

3.2.4. Predominant farming systems in the research area

The following section briefly introduces the predominant farming systems in the research area.

The total land area of Tanzania is approx. 884,000 km^2 out of which 4.5% (or 39,780 km^2) are classified as 'arable land' and only 1.1% (or 9,724 km^2) as 'permanent crop land' (THE WORLD BANK (2003)). 3.3% of the crop land is irrigated. Thus, Tanzanian agriculture is by and large rain-fed. In 2001, agriculture accounted for 45% of the Tanzanian GDP while the contribution of the sector to both, export earnings (60%) and employment (84%) was substantial. These figures underline the central role of agriculture for the Tanzanian economy. Traditional export crops include coffee, cotton, cashew nuts, tobacco, tea, sisal, and pyrethrum. However, food crop production by smallholders and subsistence farmers with average farm sizes even below three hectares characterize agriculture (URT (2006b)). The degree of mechanization in Tanzanian agriculture is rather low with roughly 70% of Tanzania's crop land being cultivated by hand hoe, 20% by ox plough and only 10% by tractor. In general there is only one tractor per 1,000 agricultural workers available (THE WORLD BANK (2003)).

According to climatic circumstances and agro-ecological conditions Tanzania has at least ten different farming systems to produce the major staple foods which comprise maize, sorghum, millet, rice, wheat, pulses (i.e., mainly beans), potatoes, bananas and plantains (URT (2006b)).

The most important farming system in the survey area is the 'maize mixed' or 'maize legume' system. These farming systems are also the most important food production systems in East Africa (DIXON et al. (2001); RUTHENBERG (1980)). They extend across the northern and middle parts of the research area with a climate varying from dry subhumid to moist subhumid. Rainfall is unimodal and erratic, however, irrigation is completely uncommon. The given farming system (maize mixed) is generally characterized by a moderately population density and rather modest average farm sizes as outlined above. Land scarcity is generally not considered to be an issue in the research area (URT (2006b)). Harvest takes place only once per year and the main staple food, maize, is mainly sold on local markets if not consumed for subsistence purposes. Often groundnuts are grown as 'cash crop' while other sources of income for the farm household are off-farm work, migrant remittances but also cattle and small ruminants (mainly goats, less sheep, see statistics in chapter 3.3.2). Cattle are kept for ploughing, milk, and breeding, but also as savings to be sold in times of scarcity. Settlement patterns are rather scattered but community institutions and local market linkages are developed to a certain extent where access roads exist. A general crisis of the maize system is apparent in the research area. Due to shortage of seeds, fertilizer and agro-chemicals, plus the high price of fertilizer relative to maize, input use has fallen in the past and continues to decline (see e.g., DIXON et al. (2001)). As a result, yields are often unsatisfying and soil fertility is deteriorating. Clearly, the main sources of vulnerability are regional droughts and market volatility (URT (2006b)).

In the dryer, semi-arid areas of the Dodoma region (e.g., southwestern part of Mpwapwa district) a more agropastoralist pearl millet/sorghum farming system can be found (URT (2006a); RUTHENBERG (1980)). The population density is rather modest, but here pressure on the limited amount of cultivated land is relatively high. The cultivation of sorghum and millet is of similar importance as livestock keeping. The cropping system is rather simple and rain-fed and resultant products are rarely marketed. Hoe cultivation is dominant in this farming system as it is in the whole research area. Sometimes land preparation is done by oxen. Livestock is mainly kept for subsistence, savings, and insurance against crop failure. The farmers in general live permanently in villages, although part of their herds may migrate seasonally under the care of herd-boys. The main source of vulnerability is drought due to poor and variable rainfall (300 to 500 mm per year), leading to crop failure and animal development below average (URT (2006a)).

In the most southern part of the research area extending into the Rubeho Mountains a partly forest based farming system can be found (700 to 1,000 mm precipitation per year). Here, farmers also practice shifting cultivation by clearing new fields from the forest. Fields are cropped for two to five years (first maize and groundnuts, subsequently cassava) and then abandoned to bush fallow for several years. With increasing population density, however, there is a general tendency to reduce the fallow periods. Maize is the main staple food, complemented by cassava, beans, vegetables, and banana. Cattle and small ruminant populations are very low and the physical isolation by the lack of roads and markets access is substantial (URT (2006a)).

After this general characterization of the research area, the next chapter will shed more light on the survey results in particular.

3.3. Case Study Results

Chapter 3.3 presents the results of the household survey in more detail, mostly by descriptive statistics. At times, inferential statistics are applied in order to test for differences between farmers participating in the TIST program and their non participating colleagues. Within the presentation of case study results, special emphasis is on the socio-economic characteristics of the interviewed households. In addition, descriptive statistics of the farming system and the livelihood strategies of the interviewees are presented. The presentation of results follows the outline used later on in structural equation modeling concentrating on the identification of the general determinants of the household decision-making process. For now it is hypothesized that those decision-making determinants can be split into six different categories, namely household preferences, the resource endowments of the household, the market incentives a household faces, as well as biophysical factors of the farm, risk and uncertainty and finally a category characterizing the social environment a household lives in. These categories are presented here with the purpose to systematize the case study results that will be further discussed later on.

3.3.1. Household Preferences

In most cases, the preferences of a household are not measured effortlessly given the limited time available in a field interview. Not seldom a certain reluctance of household members to directly tell an extraneous person their specific preferences exists. Therefore, it is controversial if preferences can be surveyed quantitatively (POATE and DAPLYN (1993)). However, if quantitative data is desired and a pre-requisite for modeling purposes, suitable proxies are needed to describe the household's preference scheme. Most often, the three main socio-demographic variables, *education*, *age* and *gender* are utilized for that purpose. The theoretical underpinning of those proxies will be discussed in chapter 4.3 (PATTANAYAK et al. (2003)).

At first, the age distribution of the interviewed farmers shall be discussed here. The age of household heads that took part in the survey ranged from 18 to 74 years with the mean age being 40 years for all households.

Table 3-1 Age Distribution of Household Heads

age	All HHs	Non-TIST	TIST
Mean	40	40	41
Median	38	37	39
Std Deviation	13	12	13
Minimum	18	20	18
Maximum	74	70	74
Skewness	,633	n.a.	n.a.
Kurtosis	-,331	n.a.	n.a.
N	200	100	100

Table 3-1 shows the age distribution of all households to approximate a normal distribution. Particular measures of that distribution, i.e., skewness and kurtosis indicating how much a given distribution varies from a normal distribution, do not reveal a great asymmetry of the data. Distributional aspects are analyzed here as they are of interest for the structural equation modeling presented in chapter 5. The skewness value of .633 indicates a slightly positive skewed distribution. However, only a skewness value greater than one would indicate a distribution that differs significantly from a normal, symmetric distribution. The Kurtosis of -.331 reveals a slightly flattened age distribution for the farm households (compared to the normal curve). In addition, the age distribution is homogenous for both, TIST and non-TIST members. The two groups do not differ significantly from each other with respect to the age of the participating household heads (t-test for equality of means: Sig. = .582). The bar chart of Fig. 3-2 visualizes the age distribution amongst participating and non-participating farmers in age classes of ten years.

Fig. 3-2 Age Class Distribution of Interviewed Farm Households

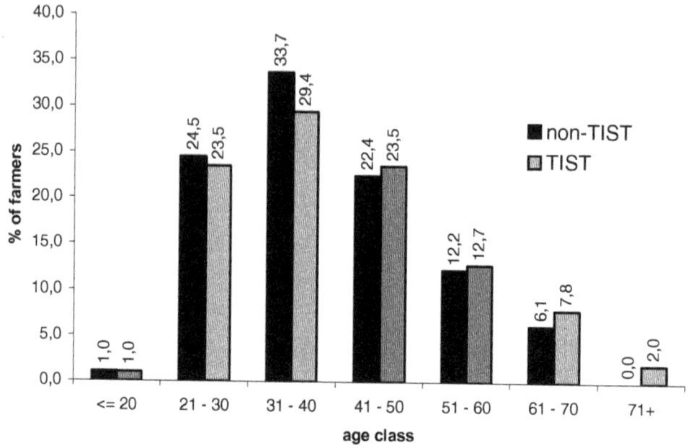

In both groups, for TIST and for non-TIST members, the age class between 31 and 40 years is most often represented. The majority of household heads is younger than 40 years which corresponds with WHO (2006) data on life expectancy at birth in Tanzania which is 47 and 49 years for men and women respectively. Fig. 3-2 shows that the age of the household head amongst TIST participants is slightly higher than for non-participants.

Fig. 3-3 Educational Status of Interviewed Farm Households

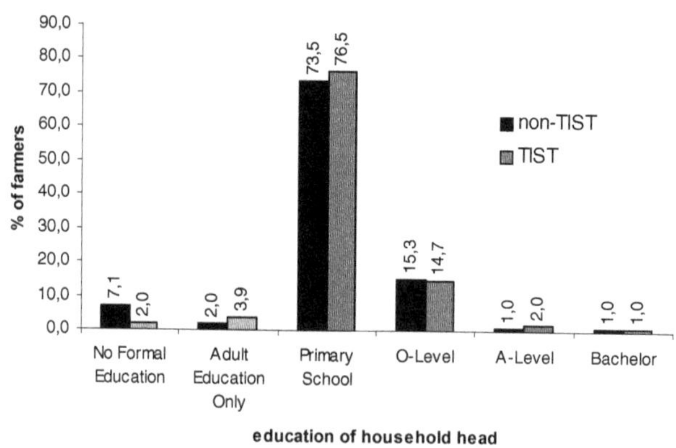

With regard to education, Fig. 3-3 illustrates that the huge majority of the farmers has a primary school education with seven years of schooling on the average. The formal education and training system in Tanzania is based on the British schooling system with seven years of primary education, four years of junior secondary (O-level), two years of senior secondary (A-Level) and up to three or more years of tertiary education. Thus, the education system has three levels, namely: basic, secondary, and tertiary levels. On average 4.6% of all farmers have no formal education at all while this figure varies slightly for TIST and non-TIST members (see chart for "No Formal Education"). Most farmers leave school after primary education, which results in only 17% of farmers that have a secondary education (O– and A– level) and even less farmers with higher degrees. For modeling purposes, the school levels were translated into years of schooling in order to obtain a metric scale variable. Table 3-2 shows the distributional properties of this metric variable. The mean education level here is accordingly seven years.

Table 3-2 Education of Household Head in Years of Schooling

educ	All HHs	Non - TIST	TIST
Mean	7,0	6,8	7,1
Median	7,0	7,0	7,0
Std. Deviation	2,7	3,0	2,4
Minimum	0,0	0,0	0,0
Maximum	16,0	15,0	16,0
Skewness	-0,181	n.a.	n.a.
Kurtosis	1,997	n.a.	n.a.
N	200	100	100

The kurtosis value (1.997) of the distribution curve for education over all households points to the fact that the distribution is more peaked than the normal curve. Non-parametric testing for differences in the two education distributions for TIST and non-TIST members did not reveal any significant findings (Mann-Whitney U test: Sig. = .321). Thus, educational achievement is not significantly different for the two groups of TIST and non-TIST members.

As will be discussed in chapter 4 the gender aspect is also supposed to influence the decision making process of farm households and finally the potential adoption of new technologies. Fig. 3-4 depicts the percentages of male and female-headed households amongst all interviewed farmers and within both groups of farmers, namely TIST and non-TIST members.

Interestingly, female-headed households within the group of TIST farmers outweigh those within non-TIST farm households by about 10%. The chi-square measure testing the hypothesis that the row (gender of household head) and column (TIST member or non-member) variables in the respective cross-tabulation are independent shows a low significance value of $p = .030$. Typically, a value below .05 indicates that there is a statistical relationship between the two variables gender of the household head and TIST membership.

Fig. 3-4 Gender Aspect Amongst Interviewed Households

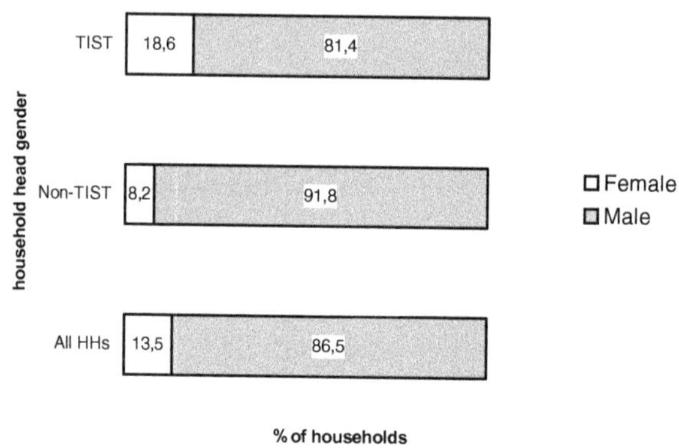

A possible gender bias within the household is hypothesized to determine the intra-household decision making process and thus finally to impact on adoption decisions of the household. Yet, the chi-square measure of .030 does not indicate the strength or direction of the relationship. Therefore, nominal directional measures indicating both, the strength and significance of the relationship between the row and column variables of a cross-tabulation were computed. Nominal directional measures are appropriate when both variables are nominal, categorical variables, which is the case here. The two nominal directional measures computed, the Goodman and Kruskal tau (Sig. = .031) as well as the Uncertainty Coefficient (Sig. = .028) indicate that there is a relationship between the two variables. Their low values of .023 for the Goodman and Kruskal tau and .022 for the Uncertainty Coefficient however remind us to act with caution when extrapolating those results. Overall the relationship between the two variables is rather weak (DALY et al. (1995)). Thus, female-headed households are not a priori more likely to participate in the program even though significantly more female-headed households can be found amongst TIST members.

In parallel to the schooling levels that were translated into 'years of schooling', the dichotomous gender variable in the following is expressed as proportion of all male household members per household in order to obtain a metric variable for the modeling. Table 3-3 shows the distributional properties of the new variable measuring the intra-household gender distribution.

The mean male household member proportion is 49% and this is pretty similar for TIST and non-TIST farm households. The distributions are slightly positive skewed and slightly flattened (negative kurtosis value). From this point of view there is no reason to suppose a major difference between the two groups of TIST and non-TIST members for the gender percentage variable.

Table 3-3 Percentage of Male Household Members per Farm Household

gepe	All HHs	Non-TIST	TIST
Mean	49,1	49,7	48,5
Median	50,0	50,0	50,0
Std. Deviation	18,3	17,8	18,8
Minimum	0,0	14,3	0,0
Maximum	100,0	100,0	100,0
Skewness	0,263	0,331	0,221
Kurtosis	-0,069	-0,116	-0,007
N	200	100	100

3.3.2. *Resource Endowments*

In order to assess the overall resource endowments of the interviewed farmers, a list of 'wealth-representing' farm household items was asked and valued at local market prices. In addition, livestock holdings were measured and valued at current market prices. The household assets asked for are either necessary for agricultural production (e.g., animal pulled plow, ox cart, hoes, axes, shovels, etc.) or for daily life (e.g., bed, chair, table, mattress, etc.). In addition, the farmers were asked for means of transportation they possess (e.g., truck, car, motorbike, bicycle, etc.), as these are specifically representative of a certain wealth status. For example, the existence of a bicycle in the household is a first indication for independence of the household in terms of selling and buying products on the closest agricultural market. Means of transport better connect the farmer to his or her environment and enable the exchange of information. This exchange will be of particular interest further down with regard to technology adoption. Fig. 3-5 shows a box-and-whiskers-plot for the sum of all household assets valued at marked prices and the value of livestock holdings for the two sub-groups of TIST and non-TIST members.

The spread of asset value for both, household and livestock assets is substantial (see exponential scale of the y-axis in Fig. 3-5). With regard to household assets, the lowest five of all households were in the range of 5,000 to 12,500 Tanzanian Shillings (TZS) which is approx. US$ 4 to 10 in wealth in household assets, while the five asset richest households were between approx. US$ 1,000 and almost US$ 6,000 (which is equal to 1,272,000 TZS and 7,409,100 TZS respectively). The asset-richest households owned fully furnished homes, three of them a car or pickup truck for transportation, and a number of different tools for agricultural production. The asset-poorest households in contrary had to manage agricultural production with only one hand hoe per household and under very poor living conditions (no bed, no chairs, no mattresses owned by the household).

Fig 3-5 suggests that the spread in asset wealth amongst non-TIST members (left side plots of the figure) is greater than amongst farmers participating in TIST.

Fig. 3-5 Summary Asset Value of Household and Livestock Assets in TZS

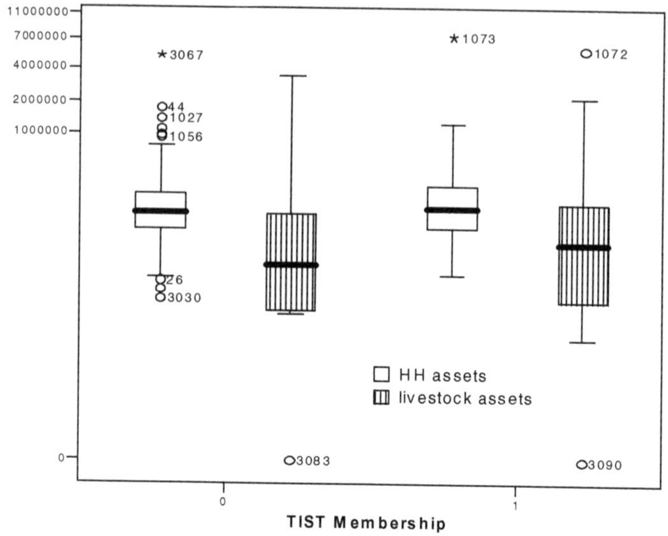

A number of outliers can be found on the left hand side in the box plot for household assets of non-TIST members (symbol for outliers °) and one extreme value lying more than three times away from the interquartile range of the box plot (symbol *). The distribution of household assets on the right (for TIST members) has one extreme value, which is marked by its household ID (1073).[28] With regard to livestock holdings the situation is as follows: 80% of all farmers own livestock and this figure does not differ greatly for TIST (81% livestock possession) and non-TIST members (78% livestock possession). Most often farm households additionally own small domestic animals, i.e., poultry (86% of all households, see Fig. 3-6). Goats (39% of all households) and pigs (26%) are common while cattle (14%), mules (9%), and sheep (8%) are more seldom owned. As outlined above, livestock possession and especially the ownership of cattle varies in the research area due to different farming systems and geographical circumstances.[29] Fig. 3-5 even shows one household in each category (TIST members and non-TIST members) that does not possess any livestock at all.

28 For the purpose of anonymization all interviewed farm households were labeled during field research with a randomly chosen two to four digit number, the so-called household ID.

29 In the flatter northern part of the research area livestock possession is more frequent than in the mountainous southern part of the area.

Fig. 3-6 Livestock of Farm Households

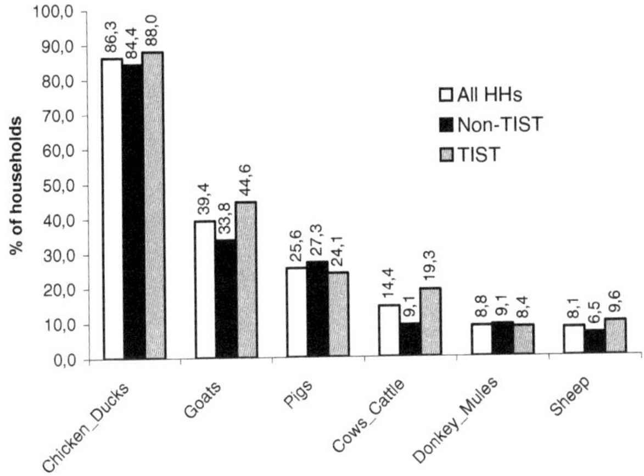

Looking at the bar chart of Fig. 3-6 in more detail, one could think that TIST farmers tend to have more livestock than non-TIST farmers (see for example the bars for 'Cows_Cattle' and for 'Goats'). However, the respective tests for this observed phenomenon (chi-square of cross-tables between TIST membership and ownership of the respective livestock in absolute livestock numbers) are not significant for any of the cases.

In a next step the distributions of the two scale variables 'household assets' (*a_hh*) and livestock assets (*a_li*) should be looked at in more detail. Table 3-4 presents the key moments of the sample distributions.

Table 3-4 Household Asset Value Distribution in TZS

	HH assets			Livestock assets		
	All HHs	Non-TIST	TIST	All HHs	Non-TIST	TIST
Mean	284630	255178	312927	228653	175112	280093
Median	132000	118750	137250	31500	19000	42750
Std. Deviation	677784,362	571678,653	767863,551	616457,007	482026,583	721213,817
Percentile 25	70500	70125	70875	3125	2375	3875
Percentile 75	232500	204550	261500	140125	115500	159000
Minimum	4400	4400	12500	0	0	0
Maximum	7409100	5124000	7409100	5880000	3458000	5880000
Skewness	7,898	6,848	8,045	5,502	4,747	5,355
Kurtosis	73,921	55,462	73,650	39,872	25,765	36,652
N	200	100	100	200	100	100

For the household assets the mean deviates substantially from the median in all three distributions, and the skewness and kurtosis values are extremely high. Therefore, non-parametric tests were used to test for potential distributional differences. However, applying the Mann-Whitney U test and the Median test, no significant difference in the locations of the household asset distributions for TIST and non-TIST members can be found (Sig. = .280). Thus, household asset endowments do not differ significantly for TIST and non-TIST farmers. Also for livestock assets, the mean deviates to a great extent from the median resulting in high skewness and kurtosis values. The data is not normally distributed and again non-parametric tests are used to detect potential distributional differences. Fig. 3-5 already indicates that the median of livestock assets of TIST farmers is higher than the one for non-TIST members. Applying the Median Test in order to detect differences in location and shape between the two groups, a significant difference of the median (Sig. = .034) can be found. Thus, the asset value of livestock endowment differs significantly for TIST and non-TIST farmers with respect to the median. This supports the intuition of Fig. 3-6 that TIST farmers tend to be better endowed with regard to livestock than non-TIST farmers as a first major finding.

A third metric variable with regard to the household's resource endowments is the sum of the household's general expenditures (*s_exp*) in the last twelve months. This value is taken as a proxy for income as it is more complicated and often very sensitive to directly ask farmers for their actual gross income.[30] Generally, the expenditure proxy works very well (see MORRIS et al. (2000)). From a theoretical point of view both, income as well as expenditures can be defended as approximations to utility (DEATON and GROSH (2000)). Thus, expenditures were taken as income proxy during the field survey. The expenditure variable includes all household disbursements except for food expenditures because food is produced in most of the cases in subsistence on the farm. The expenditures monitored comprise medicine and health costs, expenditures for clothing and shoes, festivities, transport and communication, rent of agricultural equipment or land, and expenditures for external hired labor. In addition, other farm related expenditures for seeds, tools etc., are included. Furthermore, various household consumables, e.g. soap, kerosene etc., and costs for education (school fees, uniforms, books etc.) are incorporated. Finally, 'other expenditures' like assistance to relatives or friends and costs for funerals are accounted for. Table 3-5 shows the distributional properties of the income distribution in all households and for TIST and non-TIST households respectively.

Again, a huge spread can be observed. Mean and Median differ substantially, Skewness and Kurtosis values are considerably high and positive which points to the fact that income is not normally distributed across farm households, in none of the groups. The spread of the income distribution is even higher in the group of farmers participating in TIST than in that of non-TIST farmers, however, it is not significant.

---

30 In contrary, income from a specific source like off-farm income can be asked more easily since this information does not disclose the full income situation of the household. Therefore, a greater willingness exists to report such a figure of partial income.

Table 3-5 Gross Household Expenditures per Year in TZS

sum_exp	All HHs	Non-TIST	TIST
Mean	429406	424141	434464
Median	283500	277250	296000
Std. Deviation	450233,9	426873,4	473648,6
Percentile 25	173000	155000	180500
Percentile 75	506250	516375	501625
Minimum	26000	26000	40500
Maximum	3609000	2242000	3609000
Skewness	3,239	2,160	4,012
Kurtosis	15,144	5,050	21,781
N	200	100	100

Fig. 3-7 depicts the expenditures of farm households on a *per day* basis. According to the proxy 71% of all households can only spend a dollar or even less per day to satisfy their daily needs. 90% of all households are below two US$ per day and only a very small fraction (2%) of the farm households can afford to spend more than four US$ per day for the items outlined above. While more TIST farmers belong to the second income class enabling them to spend up to two US$ per day, non-TIST farmers are the majority in the third and fourth income class but again differences are not significant.

Fig. 3-7 Sum of Expenditures per Household per Day in US$

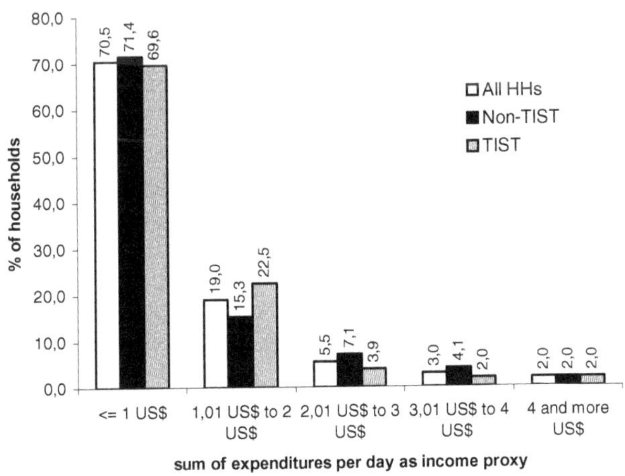

An additional variable to look at with regard to the resource endowments of the farm household is labor availability for agricultural production. In the survey, labor availability within the family was proxied by measuring household members, or descendants (*dsct*), able to assist the household head in farming activities. Those descendants, even though not yet full of age, determine the labor capacity of the household to a large extent. An estimated 3,4 million out of approx. 12 million children in Tanzania under the age of 18 work on a regular basis. In rural areas approx. 30% of all children are 'economically active' compared to approx 10% in urban areas (MASUDI et al. (2001)).[31] Thus, descendants determine labor availability in the household and are therefore putatively an important variable with regard to the adoption of a new technology. Fig 3-8 shows the distribution of descendants in all, TIST, and non-TIST households.

Fig. 3-8 Household Descendants Able to Assist in Farming Activities

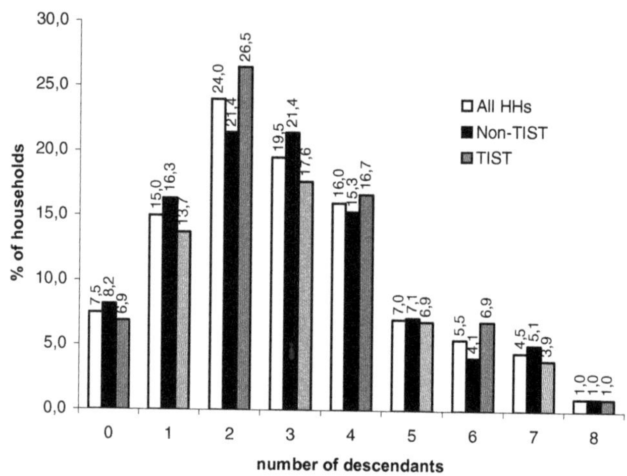

Most frequently two descendants assist the household head in farming activities (24% of all households). Approximately 8% of all households cannot fall back on own labor capacities and therefore have to either hire labor capacities or count on labor availability in the 'extended family' especially to counteract seasonal shortages.[32] Fig 3-8 reveals that there are no major differences for TIST and non-TIST farmers observable with regard to the number of descendants available to assist in farming activities.

31 Despite some progress made in recent years to eliminate child labor due to the expressed political will and a formal national commitment, the increased rural poverty during the last decade, a deterioration of the educational system and a worsening of the HIV/AIDS situation have contributed to an increase in the number of child laborers in Tanzania (MASUDI et al. (2001)).

32 In some interviewed farm households, the so-called 'extended family' was found to be the basic family unit. The 'extended family' comprises cousins, aunts, uncles, grandparents etc., aside from parents and their children.

3.3.3. *Market Incentives*

The term 'market incentives' summarizes economic determinants the farm household faces. Amongst them are prices, the availability of markets, or transportation etc., responsible for potential income losses or gains. Under this category the survey assessed an array of variables. With regard to the issue of transportation the distance to the next marketplace where to buy and sell agricultural goods (*madi*) was measured. In addition, another income related variable was assessed which measures market incentives outside the farming production system, namely off-farm income (*incm*). Moreover, the availability of credit (*cred*) to start a new agricultural business clearly determines market incentives the household faces and is therefore of interest. Finally, the price for capital (*prce*) as a general price proxy was calculated for the interviewed farm households. To begin with, off-farm income is assessed in more detail.

A farm household can realize off-farm income in several different ways through other business activities outside the own farm. 57% of all farmers were engaged in off-farm activities, but no significant difference for TIST (58%) and non-TIST (55%) farmers exists. Fig. 3-9 shows the major sources of off-farm income. Wages for employment are the most relevant source of income (38% of all farm households). Employment of this form includes for example the burning of bricks, repairing of bicycles, other small services like tailoring, etc.

Fig. 3-9 Off-Farm Income Sources of Rural Households

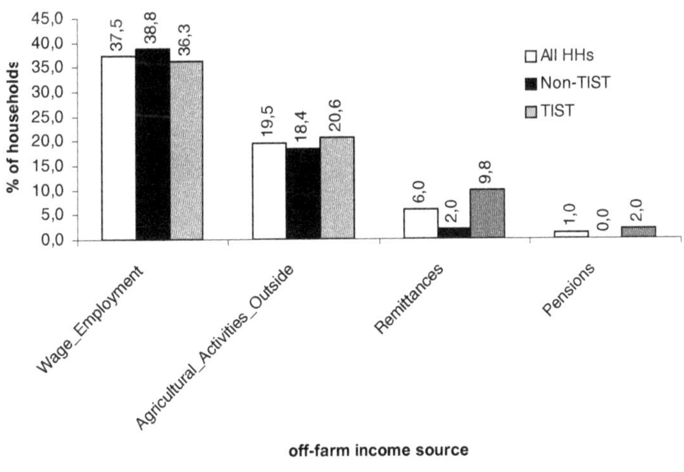

Also common are agricultural activities outside the own farm (20% of all farm households), which contribute to the overall farm income. Other forms of off-farm income like remittances (6% of all cases) or pensions (1%) are much more seldom. Interestingly, remittances do differ quite strongly for TIST (almost 10% of all cases) and non-TIST farmers (only 2%). The difference in income received due to remittances is significant at the .05 level (Sig. Pearson Chi-Square = .021). Additionally, low significance values for both, Cramer's V and the Contingency Coefficient indicate that TIST farmers receive significantly more often remittances than their non-TIST colleagues.[33]

Converting the realized income from TZS to US$ and breaking it down on a *per day* basis shows that the huge majority of the farmers (80%) is below an off-farm income of one US$ per day (see Fig. 3-10).

Fig. 3-10 Off-Farm Income Classes in US$ on a Per Day Basis

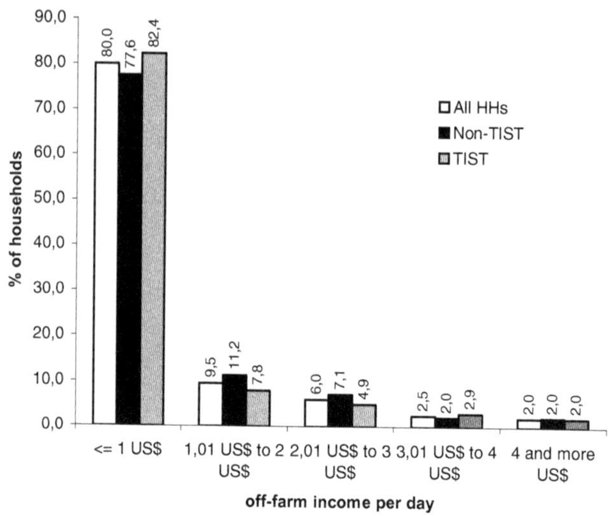

This low value is because off-farm activities are often occurring only sporadically during the year and do not guarantee a regular and reliable stream of income for the household. The observed distribution between off-farm income classes corresponds well to the observed distribution of expenditure classes in Fig. 3-7 above. Only very few farmers (2%) can realize an off-farm income greater than four US$ per day.

The off-farm income distributions for TIST and non-TIST members shown in Table 3-6 do not differ significantly for the two groups. Interestingly, the percentile 25 is zero for both distributions indicating that at least one fourth of all farmers in both groups do not realize any off-farm income at all.

33 Relatively low values for the test statistics however caution not to over-interpret the relationship (DALY et al. (1995)).

Table 3-6 Household Off-Farm Income per Year in TZS

	All HHs	Non-TIST	TIST
Mean	264725	269827	259823
Median	36000	25000	40000
Std. Deviation	525658,0	473234,9	573841,2
Percentile 25	0	0	0
Percentile 75	285000	325000	200000
Minimum	0	0	0
Maximum	3902400	2000000	3902400
Skewness	3,285	2,094	3,905
Kurtosis	14,519	3,798	19,042
N	200	100	100

The distributional moments reveal the off-farm distribution for TIST farmers to be much more spread than the one of non-TIST members. In addition, the off-farm income distribution for TIST members has more extreme values which is also shown by the mean / median deviation and the rather high skewness and kurtosis values.

With regard to credit access, Fig. 3-11 demonstrates that the vast majority of all farmers (55%) has no access to credit at all. Again, credit access differs slightly but not significantly for TIST and non-TIST members.

Fig. 3-11 Farm Household's Access to Credit

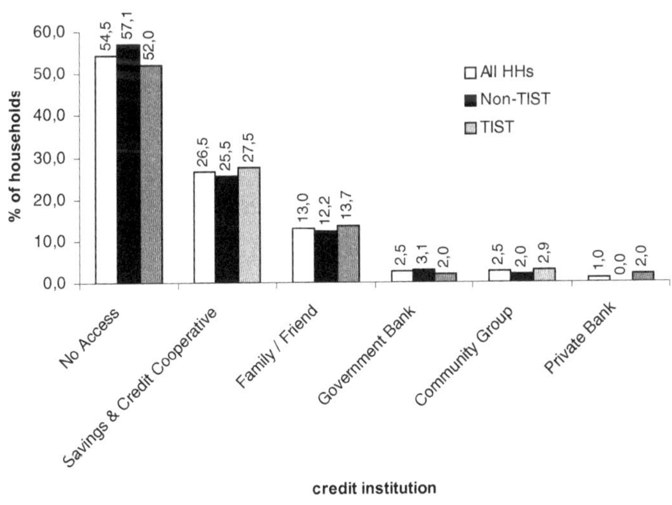

In about 27% of all cases farm households have access to credit through so called 'Savings and Credit Cooperatives' (SACCOS) that are common micro-financial organizations in rural Tanzania. SACCOS are member-driven, self-help, and not for profit institutions. The basic idea is a democratic cooperation of farmers where the members can be both, owners and users of the services available. SACCOS members save money collectively in order to grant loans to each other at often negotiable interest rates (INTERNATIONAL CO-OPERATIVE ALLIANCE (2001)). Compared to state-owned (government) and private banks, SACCOS are more popular in rural areas. One of the major reasons is their ability to provide members with emergency or consumption loans without the cumbersome bureaucratic procedures required by formal banks.

In addition, SACCOS institutions are often located in the villages of medium size, thus, frequently near their members' workplaces or homes. This considerably reduces transaction costs for farmers and makes SACCOS services more competitive than those of government banks (access rate 3%) or private banks (1%) which are often located only in the larger villages or district capitals. However, one of the shortcomings of SACCOS is that employees rarely have an appropriate education in the respective field, which often negatively affects financial and general management (IPP MEDIA (2005)). This is even more true for less formalized community groups for money lending (access approx. 3%). Thus, to establish and run a SACCOS in rural areas with personnel stemming from less educated groups is a challenging effort.

With regard to the 'informal financial sector' saving at home is the most prevalent saving mechanism in rural areas and the second best option for farmers to get credit access (13% of all cases). As this form of saving has no entry barriers and no transaction costs it is very common among the poor farm households. Saving at home enables farmers to lend money to each other, amongst peers, or even within the family. This is called 'reciprocal lending' (MUTESASIRA et al. (1999)). Under this arrangement the lender gives a credit to colleagues, relatives or neighbors based on an informal arrangement. Reciprocal lending is a common form of credit access. However, dishonesty, lack of trust and potential moral hazard explain the high average lending interest rate of up to 60% as shown in Fig. 3-12.

At a first view, the interest rates of Fig. 3-12 are astonishingly high. But this has to be put into perspective by other studies examining credit access in rural areas. HELMS and REILLE (2004) for example summarize the interest rate options available in six Asian countries in 2004.[34] The mean annual lending interest rate by commercial banks was 17.5% while microfinance institutions charged up to 80% with a mean annual lending interest rate at 48%. Informal sources of credit, e.g., moneylenders, were even reported to charge more than 100% of interest per year. Impact studies however show that clients benefit from microfinance loans even at high interest rates. An analysis in the Dominican Republic, Colombia, and Chile showed that microcredit interest costs are generally affordable compared to the income streams and total business costs of poor clients. CASTELLO et al. (1991) for example, found that even a six percent monthly interest rate represented only 0.4–3.4 percent of a microentrepreneur's total business costs.

34 Indonesia, Cambodia, Nepal, India, Philippines, and Bangladesh.

Fig. 3-12 Mean Interest Rates of Different Credit Institutions

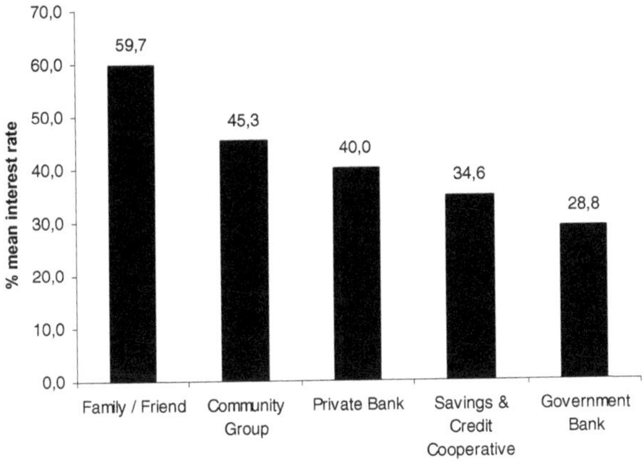

Amongst poor people access to credit is generally considered to be even more important than the actual cost of the credit (CHRISTEN et al. (1995); ROBINSON (2001)). This finding reported in the literature has been proven true for the majority of interviewees. Thus, the situation in Tanzania is not exceptional.

The different lending interest rates of the credit institutions and the access of the respective farm household allow us to calculate a variable representing the price of capital at the individual farm household. Fig. 3-13 shows the distribution of the interviewed farm households according to different lending interest rate classes by a chosen class width of 20%.

Households with no credit access at all are in interest rate class eight (140+ %). These households face a prohibitive interest rate.[35] The distribution in Fig. 3-13 suggests that TIST members might face a slightly lower cost of capital than their non-TIST colleagues. In interest rate class one (less than 20% interest rate) for example there are 7% non-TIST farmers and 13% TIST farmers while in higher interest rate classes the non-TIST farmers tend to outweigh their colleagues. Thus, there is a small but insignificant difference between the two farmer classes suggesting that non-TIST farmers have access to credit only by paying slightly higher interest rates than their colleagues.

35 Other infrastructural aspects have not been taken into account here because in each of the twenty visited villages there were at least some farmers who had access to credit. It is thus assumed that in principle all other farmers of the same village could have access to credit given the fact that collaterals are more or less the same.

Fig. 3-13 Distribution of Households According to Interest Rate Classes

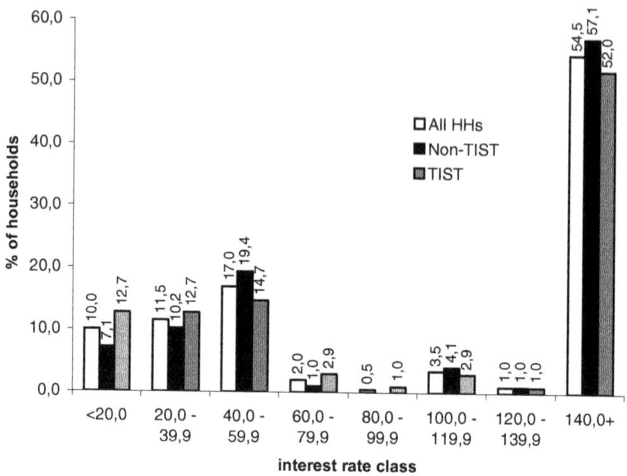

At the outset of this chapter it was already discussed that market access is influenced by available means of transportation to the market, but also by mere distance. The closer the household lives to the market the easier agricultural as well as other products can be traded. The distance factor should, especially in rural areas, not be underestimated. Roads are almost never tarred and during the rainy season in very bad conditions, littered with road holes. These conditions are often aggravated by the fact that farmers have to walk and carry all the items they intend to sell or they have bought. At best, a bicycle is available in the household. But still, transportation is rather burdensome. Thus, market distance can also be interpreted as a proxy for market incentives. In addition, the survey asked farmers how often they actually visit the nearest market. Measuring the correlation between the two variables of market distance and market visiting quantity, a significant negative correlation can be found (see Table 3-7).

Table 3-7 Nonparametric Correlations Between Market Distance and Market Visiting Quantity

			Market_Visiting_Quantity
Kendall's tau_b	Market_Distance	Correlation Coefficient	-0,250**
		Sig. (2-tailed)	0,000
		N	200
Spearman's rho	Market_Distance	Correlation Coefficient	-0,266**
		Sig. (2-tailed)	0,000
		N	200

**Correlation is significant at the 0.01 level (2-tailed)

Hence, it can be stated that the closer the farm household is to the market the more often farmers visit the market and vice versa. The correlation would have been even stronger if the Pearson correlation coefficient would have been computed. But as Table 3-8 shows the data is not normally distributed indicating that nonparametric correlation coefficients have to be computed and analyzed.

Table 3-8 Market Distance of Households

madi	All HHs	Non-TIST	TIST
Mean	3,3	3,4	3,3
Median	1,0	0,9	1,0
Std. Deviation	6,413	6,488	6,373
Percentiles 25	0,25	0,25	0,25
Percentiles 75	5,00	5,00	2,88
Minimum	0	0	0
Maximum	40	40	40
Skewness	3,775	4,073	3,530
Kurtosis	17,001	19,939	14,909
N	200	100	100

Table 3-8 shows that the average farmer resides approx. three kilometers away from the next market place. This value differs for non-TIST and TIST farmers as the respective distributions do as a whole. Looking at the distributional moments an interesting observation is that TIST participants seem to live nearer to the next market than non-TIST participants (e.g., percentile 75 for non-TIST farmers: 5 km vs. percentile 75 for TIST farmers only 2.88 km). However, the differences revealed are not significant, i.e., the distributions differ neither significantly in shape nor in location.

3.3.4. *Biophysical Factors*

The biophysical farm environment can be characterized by several different variables. All these variables jointly describe resources of the farm that are crucial for the physical production process. Implicitly, these variables are also economic determinants of the production process as they affect costs and returns (MERCER and PATTANAYAK (2003)). Of particular interest with regard to the modeling later are a) soil quality of the farm and b) slope of the farm plots as well as c) size of the farm in general. Although ambiguous views exist in the literature, as will be discussed in chapter 4, poorer biophysical farm characteristics are in general supposed to positively impact on the decision of the household to adopt a new technology (PATTANAYAK et al. (2003)). As outlined above, the focus of this study is on small scale and mostly subsistence farmers. The Tanzanian District Integrated Agricultural Survey (DIAS) which is conducted every five years by the Tanzanian Ministry of Agriculture in cooperation with the Tanzanian National Bureau of Statistics defines large scale or commercial farms as follows (MIN. OF AGRICULTURE / NBS (2001)). Commercial or large-scale farms should comprise at least 20ha of cultivated land, or 50 herds of cattle, or 100 goats, sheep, or pigs, or should have 1,000 chicken. The greatest part of the produce should be sold on the market and the operation of the farm should be continuous. In addition, there should be use of machinery on the farm and at least one employee.

Fig. 3-14 Boxplot: Land Use Quantity of TIST and Non-TIST Farmers in ha

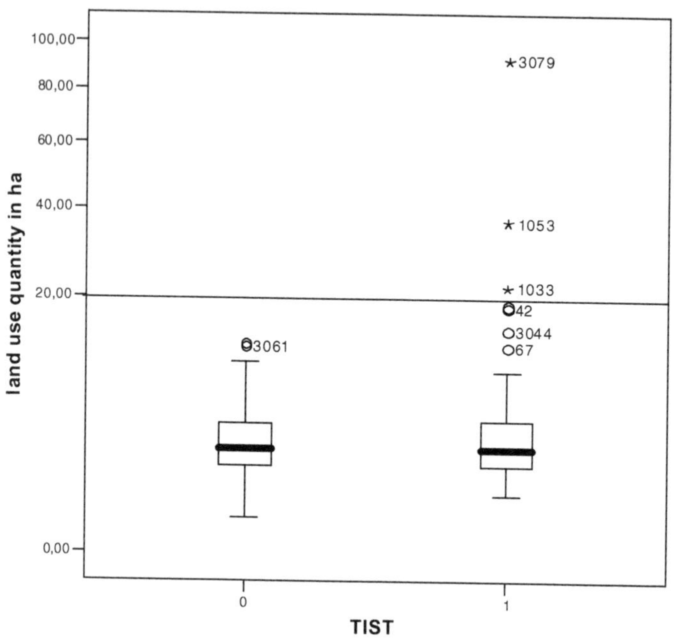

Following these definitions clearly none of the interviewed farm households was of commercial scale even though the boxplot of Fig. 3-14 shows that some very few TIST farmers overcome the threshold of 20ha farm size.

Looking into the data in more detail it becomes clear that the three households bypassing the 20ha threshold (marked by an * in the right boxplot of Fig. 3-14) do actually not meet the other criteria described above. This would not classify them as large scale or commercial farms. An interesting finding is also that household ID 3079 even though possessing almost 93ha of land is only able to use 17% (equivalent to 15.4ha) for agricultural production due to constraints in labor availability and farm equipment. Thus, even though socio-cultural circumstances privilege the household (in this case a leadership position in the village for several generations that led to greater wealth accumulation), agricultural productivity cannot be increased accordingly.

Considering the aforementioned constraints, data on farm size can be adjusted accordingly for all households to measure land holdings the household is actually able to use for agricultural production. Fig. 3-15 shows the distribution of households according to their farm size exploited for agricultural production.

Fig. 3-15 Household Distribution According to Farm Size Used for Agricultural Production

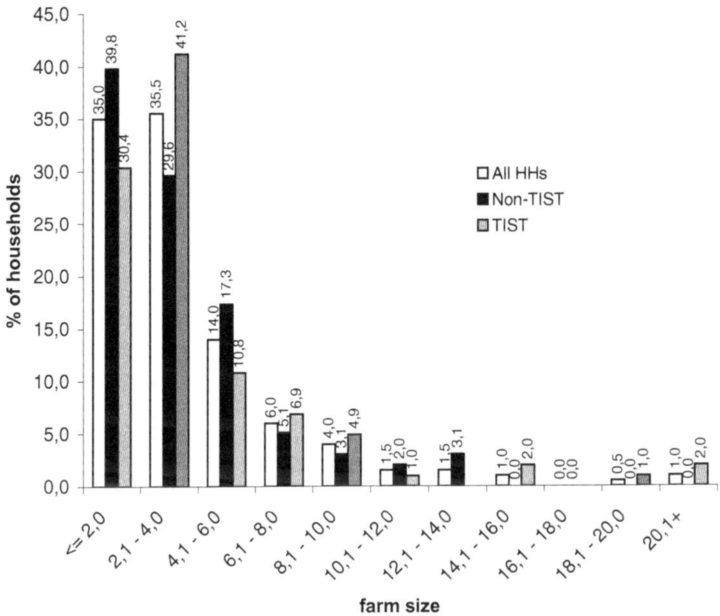

It becomes evident that the huge majority of all households (71%) works on farm sizes below four hectares and more than 90% of all farmers operate on less than eight hectares. Presumably, this fact is a challenge with regard to the adoption of larger scale afforestation activities.

Only an insignificant minority of farm households operates on farm sizes greater than ten hectares while in higher farm size classes (14ha and above) only TIST farmers can be found. The tests for significant differences between the two farm size distributions for TIST and non-TIST farmers point to the fact that overall the two groups have no different locations (i.e., moments).

Table 3-9 displays the three different distributions of land used for agricultural production with regard to all households, and the two groups of TIST and non-TIST farmers.

Table 3-9 Land for Agricultural Production (in ha)

fasi	All HHs	Non-TIST	TIST
Mean	3,5	3,2	3,8
Median	2,0	2,4	2,0
Std. Deviation	3,930	2,621	4,859
Percentile 25	1,619	1,619	1,619
Percentile 75	4,0	4,0	4,0
Minimum	0,2	0,2	0,4
Maximum	36,4	12,1	36,4
Skewness	4,305	1,780	4,117
Kurtosis	27,484	3,281	21,782
N	200	100	100

The mean size of land in use for agricultural production per household is 3.5ha while this figure varies slightly for TIST (3.8ha) and non-TIST members (3.2ha). Interestingly the percentiles for both distributions (TIST and non-TIST farmers) are exactly equal. The greater maximum farm size of 36ha in the group of the TIST farmers shifts the mean upwards in this group.

The other biophysical variables assessed in the survey were the soil quality of the respective farm plots used for agricultural production and their slope (i.e., inclination).[36] Pedologically, the research area is characterized in the northern and central parts (i.e., southwest part of the Mpwapwa district) by reddish-brown sands, in parts loamy sands. This soil type dominates in the dryer areas of the survey area with unreliable and rather low rainfall (300 to 500 mm per year). 27% of all farm households produce their goods on sandy or sandy to loamy soils (see Fig. 3-16). The major crops grown on these soil types are sorghum, pearl millet, cassava, groundnuts, sesame (simsim), and sunflowers. To some extent also sweet potatoes are grown. More to the south of the survey area dark-reddish and dark-brown loams tend to dominate. 43% of all households cultivate on loamy soils.

Maize is the major crop together with groundnuts, sorghum, sunflower, cassava, and sesame. In the more mountainous regions of the research area in the Rubeho Mountains, deep dark-reddish brown loamy clays and black-clay soils in depressions and valleys can be found. The major crops here are maize, sunflowers, beans, vegetables, and bananas. Thus, there is differentiation with regard to the soil quality across the research area. However, no clear trend or difference between the two groups of TIST and non-TIST farmers can be revealed from the data (see Fig. 3-16).

36 The surveyed plot level variables for soil quality and slope were aggregated and weighted by plot size to measure soil and slope criteria on the farm level for the purpose of the structural equation modeling in chapter 5.

Fig. 3-16 Soil Quality on Farm Level

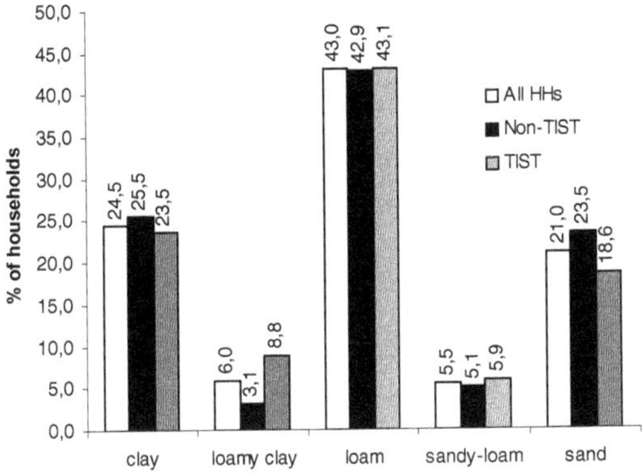

In addition, the slope of the plots used for agricultural production varies naturally according to the described geomorphology of the research area (see Fig. 3-17). The survey asked farmers to evaluate soil quality and slope based on their own farming experience. MOWO et al. (2006) demonstrated that such self-reported variables often correspond very well to more sophisticated pedological analyses. Their study compared soil quality evaluation based on experience and knowledge of smallholder farmer communities in Tanzania with the evaluation by soil scientists based on soil analysis and model calculations. Farmers' indigenous knowledge in soil fertility evaluation was to a very large extent consistent with laboratory analysis and model calculations. Thus, it is assumed here that self-reported soil quality characteristics are a good proxy for modeling purposes. With regard to the slope of their agricultural plots, farmers' appraisals were counterchecked in several cases and found to be rather un-debatable.

Fig 3-17 shows the slope of the plots used for agricultural production on a farm level perspective. 47% of all farmers produce on flat or undulating farms while 25% of all farm households operate on slightly or moderately sloped farms. Interestingly, 29% of all households produce on rather steep slopes (moderate to steep slope / steep). This finding is certainly due to the fact that plots in the more mountainous areas are favored sites with regard to yearly precipitation and thus less prone to crop loss. In addition, these sites tend to have loamy to clayey soil conditions leading to higher crop yields. Thus, the superior climate conditions outweigh the more arduous geographical setting for farming activities.

Fig. 3-17 Slope of Plots Used for Agricultural Production on Farm Level

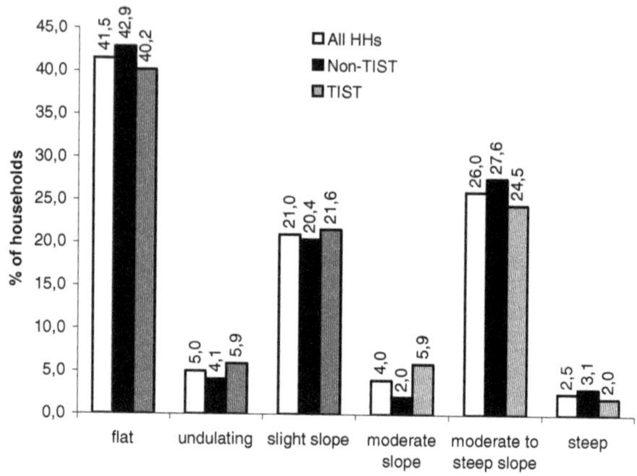

3.3.5. *Risk and Uncertainty*

Risk and uncertainty describe unknowns in the institutional and market environment under which the household has to make its decisions. The attitude of the household towards these unknown factors determines production decisions and hence resource allocation and technology choice as will be discussed in more detail in chapter 4. The chancy environment a farmer is facing is characterized by short- and longer-term risks and uncertainties. An example of longer-term risk would be tenure insecurity by missing ownership rights that restrict the usage of certain farm plots. Given relatively long gestation periods of investments in farming and especially in forestry, lower risk and uncertainty will foster technological adoption, in general (MARRA et al. (2003)). A variable for tenure restriction and two other variables described in the following were hypothesized to proxy for the 'risky' farm environment. One of them is 'years of residence' of the particular household in the community. Here, it was assumed that riskier environments in general make households move more often in the attempt to earn a better livelihood elsewhere. The third variable describes the overall labor availability in the extended family to counteract seasonal labor shortages in order to assure subsistence food production for the farm household in times of scarcity.[37]

With regard to tenure restriction, farm households were asked to report legal title or ownership rights of the plots they possess and to specify for how long in the future they would be able to use the plot without any restriction. Fig. 3-18 shows the proportion of all, as well as TIST and non-TIST farm households that were facing tenure restrictions.

37 For a definition of the term 'extended family', see footnote 32.

Overall, 15% of all farm households are restricted in the management of their farm plots. This figure differs significantly for TIST (approx. 8% of TIST households with tenure restrictions) and non-TIST households (22% with tenure restrictions; Pearson Chi-Square Sig. = .004). In addition, the nominal directional measures indicating both, the strength and significance of the relationship between two nominal variables (see explanation for the gender variable above) reveal a correlation between tenure restriction and participation in the TIST program (Goodman and Kruskal tau (value = .042) significant at .004 and Uncertainty Coefficient (value = .051) significant at .003). Thus, farm households with less tenure restrictions are significantly more frequent amongst TIST farmers.

Fig. 3-18 Restriction of Tenure Rights of Interviewed Farm Households

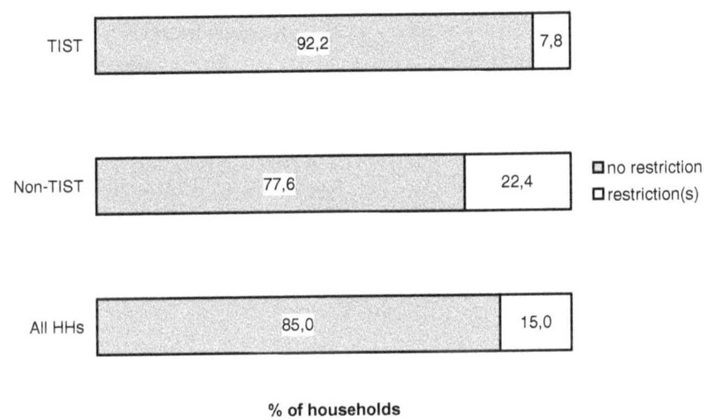

In order to obtain a variable with higher information content for the structural equation modeling presented in chapter 5, the size of all plots per farm facing any form of usage constraint was related to the whole farm size. The new variable is a proportional tenure restriction measure. Table 3-10 shows the distributional properties of this metric variable. On average 14% of the whole farm size of all households was restricted in usage be it by means of formal or informal (e.g., customary) arrangements.

All three distributions are positively skewed but differ substantially in their kurtosis values. In order to test for differences in the TIST (kurtosis value = 13.8) and non-TIST (-0.2) distributions a non-parametric Mann-Whitney U test revealed significantly different locations of the two distributions (Sig. = .003). Thus, the aforementioned finding that non-TIST farmers face significantly more often usage restrictions on their farms is reinforced.

Table 3-10 Proportion of Restricted Farm Plots

rest	All HHs	Non-TIST	TIST
Mean	0,14	0,22	0,06
Median	0	0	0
Std. Deviation	0,343	0,419	0,223
Percentile 25	0	0	0
Percentile 75	0	0	0
Minimum	0	0	0
Maximum	1	1	1
Skewness	2,102	1,341	3,870
Kurtosis	2,491	-0,206	13,759
N	200	100	100

The second risk proxy is determined by the 'years of residence' of the particular household in the community. As mentioned above, riskier environments were assumed to force households to move more often in order to earn a livelihood somewhere else. This assumption was confirmed by an interesting finding revealed by the community questionnaire. In communities where the interviewed focus group of village leaders judged the life of the people in the respective community to be worse than ten years ago due to drought, insect infestations and other natural hazards, generally migration was significantly higher than in communities with better living conditions. In addition, the mean residence time in communities with better living conditions was 8.4 years and thus significantly higher than 4.7 years mean residence time for communities with unfavorable living conditions.

Fig. 3-19 Settledness of Farm Households

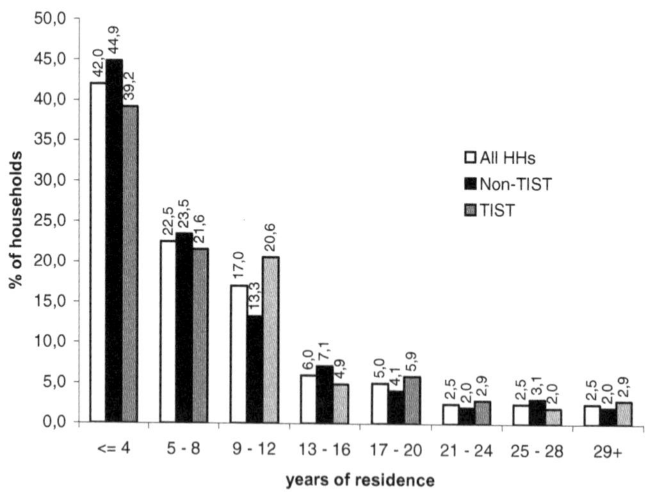

Fig. 3-19 shows a bar chart depicting household settledness in the research area in four-year-residence classes. In the lowest two residence classes (<= 4 and 5 – 8 years) non-TIST farmers outweigh their colleagues while in the third class (9 – 12 years of residence in the community) TIST farmers are more frequent. The differences are not significant, however, given the fact that the mean residence time per village is also higher for TIST farmers (8.5 years; see Table 3-11), it is reasonable to assume that TIST farmers consider their living conditions and their daily environment in general more favorable than non-TIST members.

Table 3-11 Household Residence Time Per Village

resi	All HHs	Non-TIST	TIST
Mean	8,0	7,6	8,5
Median	6	5	6,5
Std. Deviation	7,072	6,991	7,156
Percentile 25	3	3	3
Percentile 75	10	10	11
Minimum	1	1	1
Maximum	32	32	31
Skewness	1,493	1,593	1,433
Kurtosis	1,800	2,138	1,685
N	200	100	100

In analyzing the resource endowments of the farm household, the labor availability for agricultural production played a crucial role. As households mostly rely on subsistence agriculture, their food availability depends directly on the labor availability, which constitutes a clear risk with regard to the households' livelihood.

Fig. 3-20 Size of the Extended Families of the Farm Households

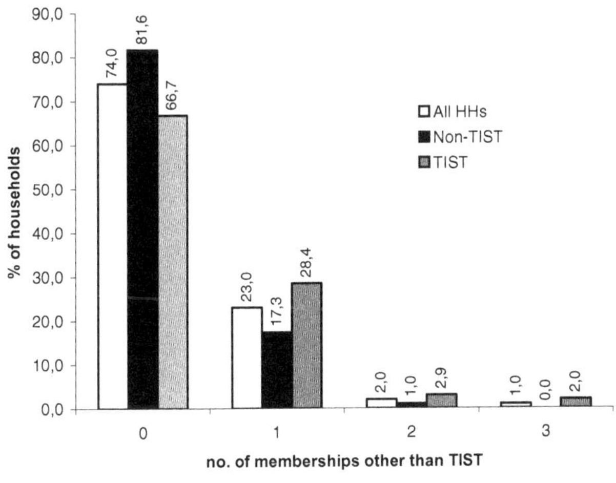

As outlined above, household descendants determine labor capacity of the household to a large extent. The analysis revealed that approximately 8% of all households can not fall back on own labor capacities but have to either hire labor or count on labor availability in the so called 'extended family' to counteract seasonal labor shortages. The extended family which comprises apart from parents and their children also cousins, aunts, uncles, and grandparents thus becomes a 'safety-net' for the nutritional needs of the farm household.

Fig. 3-20 shows that the extended families in the research area comprised up to twelve relatives. No farm household was completely dependent on own labor resources, however 9% of all households could only rely on one or two more family members 'from outside' to counteract seasonal labor shortages.

Table 3-12 Extended Family Size

flbr	All HHs	Non-TIST	TIST
Mean	5	5	5
Median	5	5	5
Std. Deviation	2,141	2,142	2,148
Percentile 25	4	3	4
Percentile 75	6	6	6
Minimum	1	1	1
Maximum	12	12	11
Skewness	0,654	0,764	0,562
Kurtosis	0,087	0,330	-0,049
N	200	100	100

Table 3-12 outlines the distributional properties of the variable describing labor availability in the extended family. Fig 3-20 and Table 3-12 show no major differences between TIST and non-TIST farmers with regard to the extended family.

3.3.6. Social Capital

Social capital is a collective term for characteristics of local social structures that affect relations among people and thus can affect decision-making and productive processes. These social structures are characterized by participatory social affiliations amongst farmers in a community. WOOLCOCK (1998) defines social capital as *"norms and networks that facilitate collective action"*. Consequently, social capital is an input or argument of the production and/or utility function of the farm household (SCHIFF (1992)). The survey questionnaire asked for several elements hypothesized to represent the social environment of the farm household. With a bundle of variables the attempt was made to measure the self-perceived connectedness of the household within the respective village community. In the following, four different variables are characterized in more detail: a) Farmer-to-farmer contacts which are measured as memberships in different organizations other than the TIST program, b) the actual frequency of those contacts as visits per year, c) farmer-to-extension contacts per year measuring availability of extension services and d) a variable assessing all additional information channels used by the household to receive agricultural market and price information. The latter includes newspapers, radio, and conversation with farmer leaders etc.

Fig. 3-21 shows farmers' memberships in different community organizations like cooperatives, associations, church groups etc.

Fig. 3-21 Farmer-to-Farmer Contacts: Memberships in Community Organizations

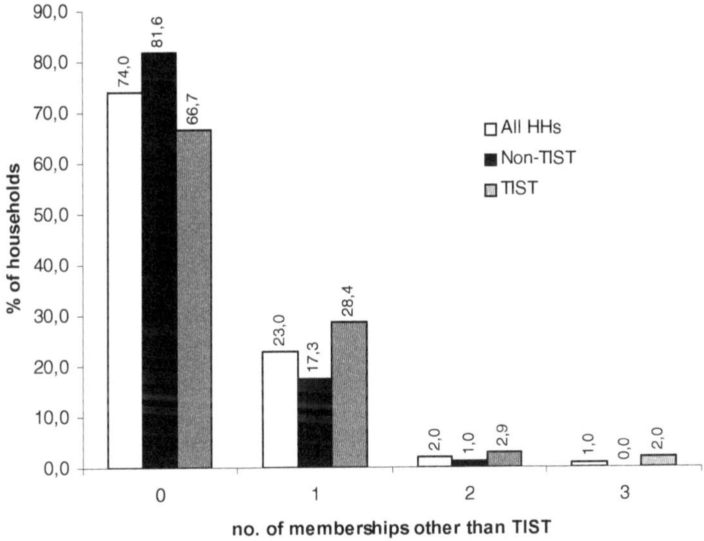

The figure displays the number of memberships besides the TIST program. Only a cumulative 26% of all farm households are members in a community organization. The majority of households (74%) are not organized in any particular form.

However, the 'rate of unionization' varies significantly between TIST and non-TIST farm households (Sig. Pearson Chi-Square = .016).[38] While only 18% of non-TIST farmers are members of any other community organization as much as one third of TIST farmers are. If at all, most of the farmers are only member in one additional organization (23%) while two (2%) or even three (1%) memberships are pretty much uncommon.

Interestingly, the actual frequency of farmer-to-farmer contacts through events organized by those institutions seems to be rather limited. Fig. 3-22 re-

38 Goodman and Kruskal tau (value = .029) significant at .016 and Uncertainty Coefficient (value = .026) significant at .015.

veals the majority of all farm households (77%) to visit the respective organizations at the most once per year. However, how many other informal contacts these memberships might generate is actually hard to quantify. There is no denying the fact that a membership in such a community organization is often a starting point for a huge variety of informal contacts that are not easily represented by the survey.

Fig. 3-22 Yearly Visits of Farm Households in Membership Organizations

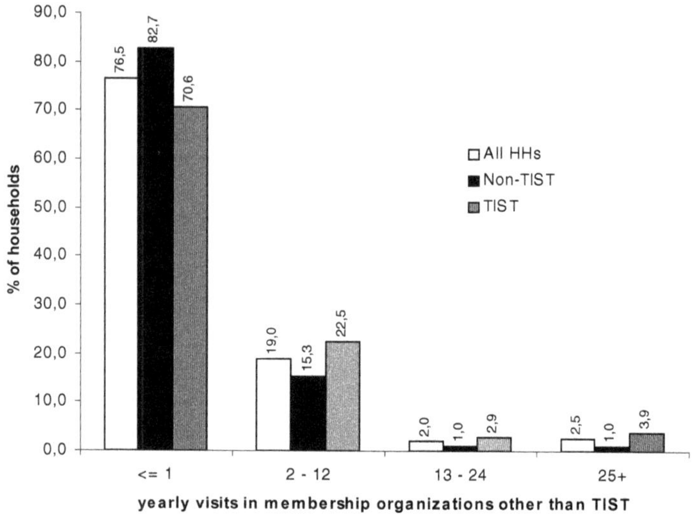

Another interesting finding is that TIST member farmers do not only belong to significantly more community organizations but also visit those organizations significantly more often during the year (Sig. Pearson Chi-Square = .044).[39] Overall 29% of TIST households visit the organizations they belong to more than once per year while only 17% of non-TIST farmers do the same.

A third variable suitable to measure the connectedness of the farm household with the outside world is a) the availability of training and extension services to the household and b) the number of contacts with those services during one agricultural year. Again, TIST farmers significantly more often receive those services than their non-TIST colleagues (Sig. Pearson Chi-Square = .022).[40] Fig. 3-23 shows the number of contacts per household in the agricultural year 2004. It becomes obvious that regular visits are uncommon but if at all

39 Goodman and Kruskal tau (value = .020) significant at .045 and Uncertainty Coefficient (value = .019) significant at .043.

40 Goodman and Kruskal tau (value = .026) significant at .022 and Uncertainty Coefficient (value = .029) significant at .021.

a regular pattern is evident it is clear that only TIST farmers are regularly visited (count of visits > 3).

Fig. 3-23 Training and Extension Contacts per Household in the agricultural year 2004

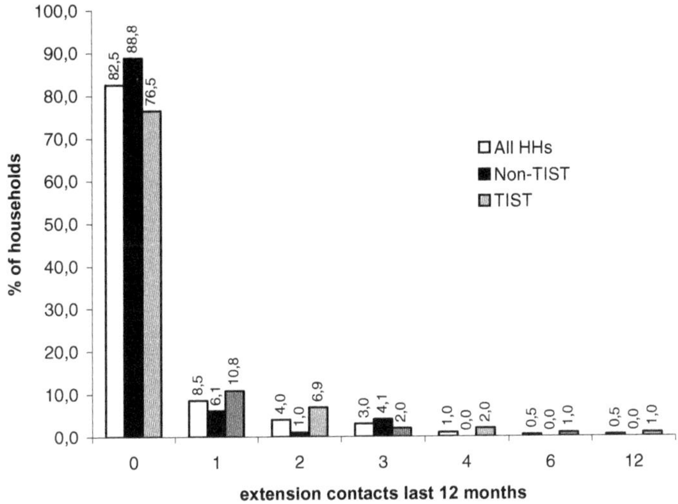

At this point it is important to note, however, that no one-to-one relationship can be made between the availability of extension services and the participation in the TIST program. This is because the Tanzanian extension service in the research area is not specifically promoting the program. The fact becomes obvious when farmers are asked for the extension advice they have received during the past agricultural year. Table 3-13 shows the subject areas that farmers discuss most frequently with an extension agent. The questionnaire also explicitly asked for tree planting / forestry related advice. As a major finding, farmers do not discuss the topic with the extension service agent.

Table 3-13 Most Frequent Subjects of Extension Service

Subject of Extension Advice	
New seed varieties	18,6%
Use of fertilizer	16,5%
General crop-related advices	11,3%
Animal diseases	10,3%
Crop disease	6,2%
Vaccination Services	5,2%
Pest infestation	4,1%
Soil problems	4,1%
Animal feed/nutrition	4,1%
Animal insemination service	1,0%

Thus, the availability of those services was taken as a general proxy for the connectedness of the farm household with its environment but no clear-cut relationship can be made with regard to the availability of extension services and TIST program participation.

The last variable presented here measures both, farmer-to-farmer as well as farmer-to-extension contacts and additional sources of information available to the farm household.

Fig. 3-24 Sources of Agricultural Market and Price Information

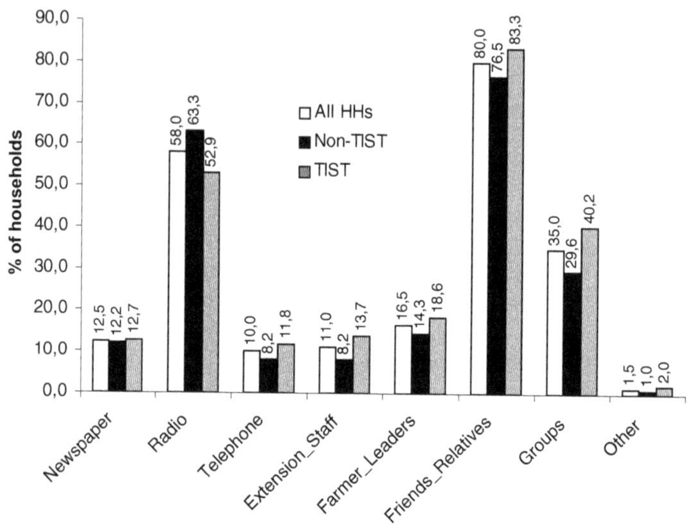

The aim was to assess which sources or channels the households use to get information about the agricultural market but also more specifically about prices of different agricultural products. Fig 3-24 reveals friends and relatives to be by far most important for agricultural market and price information for both, TIST as well as non-TIST households. 83% of TIST households for example consider their friends and/or relatives as important source for this kind of information.

An interesting aspect is the important role the radio plays in order to get more authoritative information about agricultural markets and prices. It is used by 58% of the households on average. Other sources for reliable information are extension services (11%) as well as local newspapers (13% usage for price information), and mobile telephone communication to peers living in larger villages or towns where markets are more developed (10%). Locally resident farmer leaders are considered to be even somewhat more important than the three aforementioned information sources. Again, it can be said that the various community groups (i.e., groups other than TIST) mentioned above are more important for TIST farmers than for their non-TIST colleagues.

Fig 3-25 displays a summary count of the number of different information channels for market and price information used by the households in the research area.

Fig. 3-25 Number of Different Channels Used for Agricultural Market and Price Information

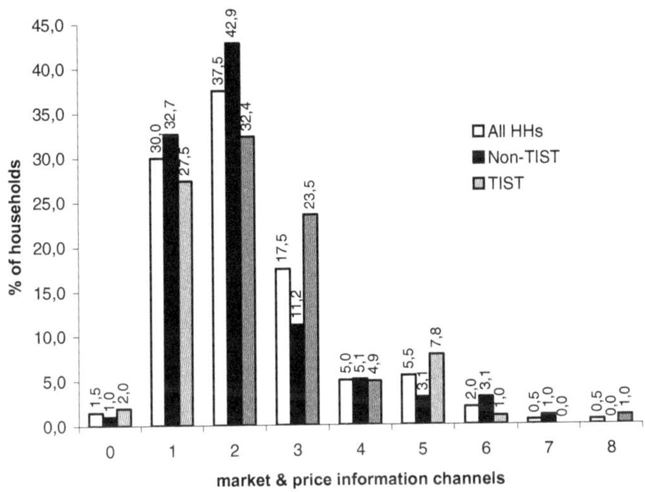

The difference in the distributions between TIST and non-TIST farm households is not significant for this variable. With higher frequencies for one, two, six, and seven information channels used it can be seen that non-TIST farm households are in general neither more remote nor more isolated than TIST farmers in terms of information access and use.

4. The Adoption of Carbon Farming – Modeling Farmers' Decisions

Chapter 4 elaborates under which conditions local farmers will most likely participate in carbon finance projects. The analysis of aspects determining participation is a necessary first step in the study of potential project impacts. Since LULUCF projects in the CDM as well as PES projects are recent, no specific literature on participation decisions exists so far. But literature on participation / adoption in other rural programs (e.g., the introduction of new agricultural technologies) can provide insights on parallels. Therefore, some results of a broader literature review on agricultural technology adoption are presented. This is presented subsequent to a brief introduction on the theory of adoption. In chapter 4 we finally develop a more specific theoretical model for the adoption of carbon farming that is based on a) the reviewed body of literature and b) the specific situation of the case study presented in chapter 3. At the end of the chapter, distinct factors determining participation are defined, which are leading to the subject of chapter 5, namely structural equation modeling.

4.1. Agroforestry Adoption – Theory and Literature Review

Before in subsequent chapters the adoption decision, or more carefully expressed the 'inclination to participate' in the TIST program, will be modeled, based on empirical data, some prior thoughts are useful: a closer look on adoption theory in general and a review of the most recent literature on agroforestry adoption. The reason for a review of agroforestry adoption literature is the following: PAGIOLA et al. (2005) point out that it is too early for a specific body of literature to exist on participation decisions with regard to PES schemes and a fortiori for such literature with regard to CDM LULUCF projects. Therefore, we decided to choose agroforestry as a technology with pretty similar features. Where necessary and meaningful for explanatory purposes other technologies, especially soil conservation were looked at as well.

Coming from an economic perspective, technology uptake is driven by three major factors (FEDER et al. (1985)):

- Utility maximization of the farmer in the context of individual and household resources,
- incentives of the external environment,
- and perceptions of the technology and its associated risks.

However, purely economic models based on profit or utility maximization fall short of taking attitudinal and social variables into account. Often these variables are equally important to explain the household decision making process with regard to adoption (ROGERS and SHOEMAKER (1971)). Social processes and structures that co-determine household's resource allocation preferences and behavior are as important as utility or profit maximizing theory. This observation involves that any attempt to describe innovation adoption behavior must combine economic *and* social aspects in a complementary way (ROGERS and SHOEMAKER (1971), PURI and NAIR

(2004), PANNELL et al. (2006)). In the following, two major explanatory approaches with regard to adoption decisions are briefly discussed, first from an economic point of view and subsequently from a more social perspective.

As aforementioned, utility or profit maximization theory is the economic starting point for the explanation of innovation adoption. Utility is usually explained in terms of return or profit resulting out of farm production and/or leisure derived from avoiding work (SADOULET and DE JANVRY (1995)). It makes intuitively and economically sense that the choice of production technology is affected by different, anticipated profits. Additionally, the household's attitudes towards risk and uncertainty determine production decisions and relate to resource allocation as well as technology choice. Furthermore, apart from the direct utility, derived from the new technology, positive external effects may be generated that influence the decision of the foresighted farmer. For instance, adoption of agroforestry and soil conservation practices is generally associated with lower rates of land clearance by present farmers, which induces positive indirect environmental effects.[41] Thus, in summary both, direct and indirect utilities influence the adoption decision of the farmer and we have to speak of expected utilities.

FEDER et al. (1985) point out that it is equally important to clearly distinguish between different levels of aggregation when modeling adoption (i.e., farm household versus aggregate adoption) and to address the aspect of an adoption timeframe. That is, adoption as a more continuous process vs. a dichotomous choice. The latter is directly linked to the so-called *divisibility* of an innovation. For a non-divisible technology, the extent of adoption at farm level is necessarily dichotomous. Farm-level adoption models frequently use such an approach. Adoption is hence modeled in terms of the likelihood that farmers with pre-defined characteristics will adopt a given technology. The new technology to be adopted can either be a complete new package of complementary technologies or only a few components. Important here is the binary choice of either one, if the technologies are non-divisible. More seldom, continuous models have been applied to describe individual farm level decision-making processes. Such models assess adoption decision making over time and are therefore able to allow for changes affecting the perception of the household with regard to the performance of the technology. Models considering aggregate adoption in principle depict the spread of an innovation with regard to a specific area and time.[42]

A basis for social explanations for adoption of innovations is the model of ROGERS and SHOEMAKER (1971). They assume that the individual runs through several stages of adoption, namely *awareness*, *evaluation*, *trial*, and finally *adoption*. The most important determinant in their model is access to information (e.g., on- and

41 A finding that is equally true for the Amazon (JONES and ONEILL (1992)) as well as for east Africa (SANCHEZ et al. (1997)).

42 Aggregate adoption models are not of primary interest for this study. For further details see FEDER et al. (1985).

off-farm). Key sources for this information are inter alia extension services and educational programs as well as friends and neighbors as seen in chapter 3. The actual time to complete the whole adoption process by an individual person is determined by specific characteristics, e.g., age, gender, and education. ROGERS and SHOEMAKER (1971) extended their initial model by describing the so-called adoption curve model, which categorizes adopters into five different groups. *Innovators, early adopters, early majority, late majority*, and *laggards*. Economists as well as sociologists have used the adoption curve model widely to develop an analytical framework to study household characteristics and socio-economic factors potentially inducing adoption. Furthermore, it is important to note that social models attempt to explain the effects of peers or community groups for adoption. The interactions between persons can eventually result in joint decisions on technology adoption and promotion of individual decisions (CONLEY and UDRY (2001)).

Social institutions that gradually change across ethnic or community boundaries also condition preferences, which in turn determine adoption decisions. The same is true for organizational structures or policies that eventually change abruptly with administrative boundaries (STAAL et al. (2002)). All these factors must in principle be taken into account when making the attempt to forecast adoption decisions.

Especially the literature on participation in forestry or agroforestry programs and adoption in developing countries provides a good starting point to gain more concrete insights on factors relevant for landowners' decision-making processes. The continuing interest of researchers and policy makers especially in the adoption of agroforestry is primarily based on the hypothesis that under specific circumstances such projects might constitute an economically preferable option to conventional agricultural systems. The former must not rely to the same extent as conventional agriculture on purchased inputs for sustained productivity (MOYO and VEEMAN (2004)). The specific advantage of agroforestry technologies requiring low external input is that particularly poorer farmers who lack capital and credit access are in need of ways to improve yields and income without expensive inputs.

The crop selection process of the farmer within an agroforestry system is in a way analogous to an investor's portfolio assortment problem. The term crop is used here in its widest sense, comprising both annual crops and perennials, which is a generic term including trees. The farmer invests the assets at his disposal (e.g., land) in the agricultural portfolio that consists of one or more annual and/or perennial crops. Some of those crops are grown for subsistence purposes while others may be determined for commercial sale. A clear-cut distinction between the two purposes is often not possible. Tree products for example are often both, consumed at home and sold on the market. As outlined before, market opportunities and household needs determine the choice and level of the activity. The actual portfolio mix between annual and perennial crops will then be influenced by economic and social factors.

SCHERR (1992) suggests that several different exogenous incentives possibly intensify the farmer's tree planting activities, e.g., declining access to wooded land whereby tree product supply is reduced; secondly, an increasing demand for tree products resulting out of either population growth or new tree uses or products, or even new markets.

Furthermore, land users may respond to declining land quality by planting or protecting trees, be it for windbreaks, erosion control, soil fertility, grazing land rehabilitation or similar purposes. Possible substitutes and other supply sources determine the rate and degree to which land users respond to such incentives. However, if land in general becomes scarce the principal need to produce food for subsistence and income overrides the long-term value of trees (ARNOLD (1983)). In addition, it is important to bear in mind that farmers generally will produce household staples on the more productive land while poorer land is theoretically targeted for the tree planting activity. Therefore, more intensive tree production is hypothesized to take place on larger farms with lower constraints on land availability (SHIVELY (1999)).

However, with the new aspect of regular carbon finance payments for the carbon sequestration service the formerly central question of the investment gestation period (i.e., 'how long can the farmer wait for the trees to mature?') must now be answered differently. It is certainly true that poverty and economic risk reduce the incentive to invest in a long term tree planting activity (SCHERR (1995b); REARDON and VOSTI (1995)). But, the carbon finance payments provide even prior to maturity a continuous stream of income, which can help to smooth scarcities.

A wide array of on the one hand more social and on the other hand more economic household and farm-related variables have been identified as determinants for participation or adoption decisions in the literature. Following the approach above, at first described, there are a couple of economic variables commonly considered essential, followed by variables in the social context.[43]

4.1.1. Economic variables determining adoption decisions

Assets

Amongst the main economic variables are *assets* describing the resource base of the farmer. The term asset subsumes various different asset types. Most importantly land holdings (i.e., size, condition etc.) followed by other assets like housing of the farmer, as well as all other kinds of durables (e.g., bed, mattress, chair table, hoes, axes, plows etc.) are relevant. In addition, vehicles play generally an important role. Vehicles are assets that indicate a certain wealth status (e.g., bicycle, motorbike, car, pickup truck). Livestock is sometimes included in the above-mentioned assets, sometimes specified separately. For agroforestry studies a separate consideration is justified given the fact that livestock is a key ele-

---

43 A precise distinction between 'economic' and 'social' variables is neither possible nor intended here. All variables can be subsumed under the broad umbrella of the so-called 'socio-economic variables'. This term indicates an array of social and economic indicators commonly used to describe living circumstances.

ment of various agroforestry systems in Sub-Saharan Africa (MOLUA (2005)). From an economic point of view, all assets under the control of the farmer possibly generate income (i.e., 'returns on assets', see chapter 2.4), e.g. the plot used for the production of crops.

The impact of asset holdings on the decision process with regard to adoption is unambiguous. The correlation is reported consistently positive indicating that greater wealth and increased availability of capital make the investment in a new technology more feasible (MERCER and PATTANAYAK (2003); PATTANAYAK et al. (2003); FEDER et al. (1985)). Furthermore, farmers that are better off are prepared to take more risks potentially associated with the investment decision than poorer farmers. This is due to the fact that better endowed farmers can afford to take greater chances (MAS-COLELL et al. (1995)).

Soil quality and slope

With regard to the land holdings of a farm household, apart from the mere size of the plot(s), often more specific characteristics like soil quality and slope are reported. Although in most cases demanding to measure, these variables further describe the most important resource of the farm household's production process, which argues for their inclusion. The correlation results with regard to adoption are mixed (ADESINA and CHIANU (2002); ZBINDEN and LEE (2005)). Typically, one would expect that poorer soil qualities, steeper slopes and with that an eventually higher risk of soil degradation and erosion are positively correlated with adoption. Furthermore, it is economically reasonable to assume that a farmer would choose to reforest marginal lands within the farm system. However, the set of variables used to measure soil characteristics (sometimes by direct investigation, sometimes rather subjective by self-reported values) delivers ambiguous results and often needs further interpretation in view of the specific situation. Some of the literature on adoption of agroforestry systems addresses irrigation as a separate variable to further characterize the production system. However, the huge majority of agroforestry systems in developing countries are rain-fed production systems, as it is the case in Tanzania where only 3.3% of the cropland is irrigated (see chapter 3.2.4). For that reason irrigating is not specifically considered in the study at hand.

Tenure

Another crucial point apart from size and biophysical characteristics of the plot are tenure related regulations. In the case of agroforestry, the correlation seems to be pretty clear. In general, tenants have a lower interest to adopt agroforestry technologies on the land they do not own (PATTANAYAK et al. (2003)). The long time horizon of the tree planting activity explains this correlation. However, findings are not unanimous. A considerable debate can be found in the literature on tenure related arrangements and adoption decisions. The reported correlations are in both directions, negative and positive (FEDER et al. (1985)).

Household income

Household income is usually measured with regard to different income sources, i.e. directly farm-related income from agricultural activities and off-farm income sources such as wages. The correlation of income with adoption is typically positive as in the case of assets. However, some ambiguities are reported. Where a farmer predominantly relies on off-farm income there might be simply no interest to adopt a new technology (KEBEDE (1992)). On the other hand, risk averse farmers who completely rely on farm-related income might also not risk the investment in a new and unknown technology (PATTANAYAK et al. (2003)).

Credit availability

The effect of credit availability on adoption decisions is equivocal as well. Firstly, credit is often not available in rural areas of developing countries. Furthermore, where credit is in principle obtainable especially smaller farmers often lack the collaterals to get access or interest rates are prohibitively high (see chapter 3.3.3). Overall, there is insufficient evidence that access to credit might either speed up smallholder farm forestry or that the lack of credit access might completely prevent its adoption (MAHAPATRA and MITCHELL (2001)). In the context of their Latin American study CURRENT et al. (1995) for example conclude that *"formal lines of credit for agroforestry may not be essential in promoting smallholder agroforestry."* The results of PATTANAYAK et al. (2003), however, point in the opposite direction. Credit availability had a statistically significant positive correlation on adoption in the reviewed studies. These contradictory findings might be explained partly by the fact that research in general has a certain predisposition to only publish significant results, which is of course equally true for the other variables outlined here.

Prices

With regard to prices, cross-sectional data has various constraints. First of all, cross-sectional price data is time invariant. Furthermore, limitations concerning geographical aspects of the data (i.e., distance to markets determining prices in rural economies) and often a lack of statistical variability, which might be due to the fact, that data is usually self-reported, come into play. Prices of products associated with the tree planting component of the agroforestry system will be rather positively correlated with adoption while prices of competing crops correlate most likely negatively (SHIVELY (1999)). The lending interest rate can be interpreted as price for capital and acts in the study at hand as a general price proxy (see further down).

Labor

Another important economic variable touched upon in chapter three is the availability of labor. Labor assessments are in general based on the number of (mostly male) adults in the family. Often also the size of the entire family including children over a certain age limit is taken as a proxy for labor availability. A high labor demand of the agroforestry technology supposedly limits a farmer's use of it. However, as AJAYI et al. (2003) show, labor constraints might not be decisive for testing the technology but for a *continuous* adoption as well as for the amount of land allocated to the new technology over time (i.e., intensity of adoption). In addition, members of poorer households are often forced to sell their labor force by working for other farmers in order to earn the daily staple food. This actually prevents the poorer farmers from participation even if they wished to do so. On top, opportunity costs of labor play a role in the adoption decision. Higher wages in rather peri-urban areas mean higher opportunity costs of labor than in remote rural areas. Given the labor intensity of forest planting activities, higher opportunity costs of labor might impede adoption. The more rural the areas the lower the opportunity costs of labor, in general. This may induce greater use of labor intensive technologies (ADESINA and CHIANU (2002)). Numerous studies have confirmed this correlation in determining program participation or technology adoption. The sign of the correlation is in most of the cases positive indicating that greater labor availability promotes agroforestry adoption (AYUK (1997); NEUPANE et al. (2002)).

4.1.2. Social variables determining adoption decision

The focus of the next paragraphs is more on the social side of the adoption decision and the variables commonly utilized to describe this interrelation. First, the three main socio-demographic variables, education, age, and gender are discussed. Followed by an analysis of both, farmer-to-extension and farmer-to-farmer contacts. Last but not least, religion and ethnicity are briefly discussed.

Education

There are usually two ways to measure the educational status of a household. Either, only the educational level of the household head or the average level of education of all household members is reported (GLEWWE (2000)). The rational for the assessment of this variable is the following. A rather knowledge and management intensive technology requires greater managerial ability of the farmer and hence advanced skills in order to achieve best possible production results. The bottom line is that a lack of knowledge can either completely prevent farmers from adoption. Or, if the farmer adopts, such a knowledge lack can result in poor performance of the new technology and finally to its abandonment (ATTA-KRAH and FRANCIS (1987)). This implies in theory that adoption of households with relatively higher educational achievements is more likely. Some of the literature supports this theory (HERATH and TAKEYA (2003); WELCH (1978)). However, in other studies educational variables were not significantly associated with an adoption decision (AJAYI et al. (2003); CAVIGLIA-HARRIS (2003)). Thus, education is a pretty standard socio-demographic variable but its correlation with adoption is sometimes rather poor (PATTANAYAK et al. (2003)). An explanation might be the somewhat limited variation in the educational data (see chapter 3.3.1). A phenomenon that is not only true for educational standards in Tanzania but for other developing countries as well. Top-down educational programs, designed to guarantee a certain minimum standard of education for the rural population, often characterize developing countries (UNESCO (2004)). This in turn leads to rather homogenous educational standards even though on a relatively low level.

Age

The effect of the farmer's age on the participation decision may theoretically also be either positive or negative. On the one hand, the age of the farmer is linked to longer farming experience and hence greater accumulation of knowledge. As in the case of education this is expected to have a positive effect on the adoption decision. On the other hand, shorter planning horizons and higher risk aversion of older farmers may have the opposite effect. In addition, MBAGA-SEMGALAWE and FOLMER (2000) report older farmers to be less equipped and less motivated than younger ones which affects adoption negatively. In contrast, the longer planning horizon of younger farmers makes them more likely to engage in investments with longer gestation periods, as is the case in tree planting activities without carbon finance revenues. The net effect of the age variable on the adoption decision is therefore hard to determine a priori (LAPAR and PANDEY (1999)).

Gender

Both, binary as well as continuous variables, cover the gender aspect of adoption in the literature. A binary variable only indicates if the household is male or female-headed, while a continuous variable most often reflects the proportion of males in the household (see chapter 3.3.1). PATTANAYAK et al. (2003) report the gender variable to be one with the highest explanatory power amongst the social variables. The correlation is positive with a higher proportion of males being more likely to adopt agroforestry technologies. However, one should act with caution when interpreting such results as they might actually reflect an existing gender-bias with regard to land allocation and/or traditional inheritance systems. Women might not or might only have restricted ownership of land. Land heritage might only be possible for men. In addition, the right to actually own trees or even plant them might be restricted (ADESINA and CHIANU (2002)). In Bangladesh for example exists a well-defined gender separation of the labor force with regard to tree growing activities (SALAM et al. (2000)). While male household members are usually responsible for the selection of planting sites, species, the collection of seedlings and planting them, female household members are rather involved in caring for the planted trees. Therefore, in such cases the number of males in the household is necessarily positively associated with an adoption decision to plant trees (LEUSCHNER and KHALEQUE (1987)).

Farmer-to-extension and farmer-to-farmer contacts

Different channels exist through which farmers may pick up information about a new technology. Besides extension services, information disseminated via the media plays a role in the adoption process. In addition, farmers may learn about the characteristics of the new technology from their neighbors and peers (CONLEY and UDRY (2001)).

Within the different stages of adoption in the model of ROGERS and SHOEMAKER (1971) (i.e., *awareness*, *evaluation*, *trial*, and *adoption*) information disseminated via the media or extension agents play a more important role in earlier stages (awareness raising) while interpersonal contacts to other farmers are more decisive for adoption in the later stages. The focus here is first on farmer-to-extension and subsequently on farmer-to-farmer contacts. Contact with or access to extension services - usually reported as binary variable - has been positively correlated with participation decisions in a number of studies (CHIANU and TSUJII (2004); GLENDINNING et al. (2001); MAHAPATRA and MITCHELL (2001)). Access to extension services usually means that a certain kind of training was received with regard to the technology under consideration. Knowledge and information about the new technology in turn improve the profitability of initial technology trials and hence reduce the number of years before adoption takes place (GHADIM and PANNELL (1999)).

The second aspect fostering adoption are farmer-to-farmer contacts which are measured as memberships in different organizations. Some studies found that agroforestry extension and development projects that work through farmer groups have higher success rates (ADESINA and CHIANU (2002); ATTA-KRAH and FRANCIS (1987)). Memberships in different organizations form a specific network that is hypothesized to mitigate risks and uncertainties related to the adoption of a new technology. It is therefore expected that households that are part of a cooperative or community organization will benefit from this network and thus quicker adopt new technologies. CAVIGLIA-HARRIS (2003) for example found that union membership is a main determinant of adoption. In their case study unions not only taught farmers about cultivating and marketing issues of crops but also assisted in price negotiations. In the study of tree planters and non planters in a subsistence farming system by MAHAPATRA and MITCHELL (2001) the main influence on uptake of tree planting was membership in village organizations. A variable measuring such different memberships should therefore positively correlate with the participation decision.

Religion and ethnicity

Finally, measures of the socio-cultural status of the household, respectively the household head must be discussed, at least briefly. Such measures include religion, ethnicity and eventually tribes or castes. In societies differentiating according to the aforementioned measures, usually majority and minority groups exist. The socio-cultural status is linked to preferences of the household. If the household belongs to an ethnicity, which is rather 'pro-adoption', this may facilitate the household's internal decision process. In the case of BEKELE and DRAKE (2003) for example, being a member of the majority ethnic group, is expected to have a positive correlation with the conservation decision. In addition, socio-cultural characteristics are interpreted as proxies for risk-taking attributes of farm households hence also determining the decision process (PATTANAYAK et al. (2003)).

4.2. A Theoretical Model for the Adoption of Carbon Farming

In the following paragraph a theoretical model of the farmer's adoption decision i.e., whether or not to participate in the TIST program, is developed. Starting point is the conceptual household model of SADOULET and DE JANVRY (1995). Their focus is on an integrated production, consumption, and work decision of a peasant household. The model developed here will take the problem one step further by differentiating between agricultural production and tree planting decisions of the household. In addition, the fact that TIST is a 'Small Group-based' tree-planting program brings interactions of the farmers and hence social capital into play. 'Social capital' is generally defined as networks and norms existing in those networks that both potentially determine collective action and decisions (WOOLCOCK (1998)). Along these lines, social capital will be taken into account and extend the initial model of SADOULET and DE JANVRY (1995).

Our farm household is seen as a representative agent aiming to maximize its utility, U. Risk aversion determines the shape of the utility function to be concave. Furthermore, the function is continuous and twice differentiable.[44] The household maximizes its utility by consuming an agricultural good, C_a, a good purchased on the market, C_m, and a certain amount of home time or 'leisure', C_l.[45] Furthermore, utility is maximized over the tree planting activity, captured by B. B provides the household for instance with firewood and fruits from the trees. B depends not merely on the number of trees planted by the Small Group

44 A concave utility function has a positive first derivative and a negative second derivative with respect to wealth. Concavity implies that each increment to wealth conveys progressively smaller increments to utility. Decision makers with concave utility functions are risk averse (PENNINGS and SMIDTS (2003)).

45 SADOULET and DE JANVRY (1995) explicitly point out that 'home time' is a more comprehensive term for time spent not directly in productive and labor market activities. The concept of home time takes other activities like family maintenance, family reproduction (i.e., pregnancies, rearing children etc.), and time for socialization with neighbors and the community into account. The bottom line is that only a small fraction of home time is actually spent for leisure (i.e., relaxation, pleasure, and sleep).

the farmer is a member of but, in addition, the 'performance' of the Small Group influences the production function of *B*. The utility function is shaped by household socio-demographics that are proxied by preferences, *H*. The maximization problem is subject to risk and uncertainty on the short- and long-run, *R*. Thus, the household's utility optimization problem is expressed as follows in equation (4.1):

$$\text{Max} \quad U((C_a, C_m, C_l, B; H), R) \qquad \text{Utility function} \quad (4.1)$$

As described by SADOULET and DE JANVRY (1995) households are 'semi-commercialized' in the sense that, even if all markets work, at least some of their agricultural production is kept for home consumption and some of their labor resources are used directly for home production. Food produced in excess of household consumption is sold on the local market, and family labor in excess of use on the home plot is sold on the labor market. The simultaneous production, consumption, and work decisions of the household imply that the utility maximization is subject to three constraints, namely,

- a technology constraint, expressed by the respective production functions,
- a cash constraint,
- and a time constraint (SINGH et al. (1986)).

The time constraint expresses the fact that the sum of own input supply of time for production (i.e., labor, L_s) and home time consumed by the household, C_l, can not exceed total time available, *T*. However, total time available of the household producing a certain agricultural good can eventually be increased if the household hires off-farm labor L_o.

$$T = L_s + C_l + L_o \qquad \text{Time constraint} \quad (4.2)$$

The overall household production capacity can be used interchangeably for the production of agricultural goods Q_a, and/or the planting of trees *B*. Hence, there is an implicit trade-off between the two production outputs. However, the production cycle of the agricultural year, in general, allows some degree of flexibility with regard to the planting of trees.

The production function of the agricultural product, Q_a, with price P_a draws on two variable factors, which are labor, L_s^a with price *W* (wage rate), and another non-labor variable factor X_a with price P_x^a. Other fixed factors such as household resource endowments, *E*, and the biophysical farm characteristics, *Z*, condition the production function that is assumed to be concave and continuous. The production function of the agricultural product constraining the utility maximization is expressed in equation 4.3:

$$Q_a = F_1(X_a, L_s^a; E, Z) \qquad \text{Production function } Q_a \quad (4.3)$$

The production function for *B* is developed analogously. In addition, the group-specific interactions, *G*, of the member farmers become relevant. Such group-specific attributes as they are perceived by the respective farmer include for example the size of the group, the distribution rule for the carbon finance payments amongst farmers, the overall 'activeness' of the group with regard to its participation in the program (e.g., did the group ever submit best practices to help improve the overall program design etc.).

$$B = F_2(X_B,\ L_s^B;\ E, Z, G) \quad \text{Production function B} \quad (4.4)$$

$$\text{with } L_s = L_s^a + L_s^B \quad (4.5)$$

Production of the agricultural good Q_a compared to its consumption, C_a, and input supply of time for production, L_s, compared to total time available, *T*, and the demand for home time, C_l, determine whether the household is a net buyer or seller of food and labor. Multiplied by the respective prices, P_a and *W*, plus the direct carbon finance payments, P_C, the price for wood, P_W, and non-timber forest products (NTFP), P_{NTFP}, multiplied by the activity level of *B*, these factors are the right hand side of the cash constraint of the peasant household. They cannot exceed the sum of the price, P_x , the household can pay for the variable production factor *X* and the price P_m for the good purchased on the market, C_m plus any inputs (like seedlings) for the tree planting activity multiplied by their respective prices.

$$P_x^a X_a + P_x^B X_B + P_m C_m = P_a(Q_a - C_a) + (P_C + P_W + P_{NTFP})B + W(T - L_s - C_l)$$

Cash constraint (4.6)

For solving the maximization with the respective constraints, the representative household faces the choice variables: C_a, C_m, C_l, Q_a, B, L_s^a, L_s^B L_o which are all functions of the exogenous parameters *R*, *E*, *Z*, *H* and market-based incentives, *I*, as explicit and implicit prices of the outputs and inputs of the production process. In case the expected net utility, expressed as difference between the scenario with and without tree planting is positive, the household will participate in the project, and the value of *B* will be greater than zero. A specific activity level of *B* is not examined here. Doing so would mean to assess the intensity of adoption. For the problem at hand only the 'corner solution', i.e., *B* being zero or the information that *B* is greater than zero is of interest.

4.3. Factor Categories Determining Participation

This paragraph further develops the idea already introduced in chapter 3. The case study results in chapter 3 were grouped into six different categories, namely a) household preferences, b) the resource endowments of the household, c) market incentives a household faces, as well as d) biophysical factors of the farm, e) risk and uncertainty and finally a category characterizing f) the social environment a household lives in. Chapter 4.3 elaborates on the hypothesis that the grouping of descriptive variables for adoption unfolds a greater explanatory power with regard to the household's adoption decision. Thus, the underlying hypotheses for the structural equation modeling of chapter 5 are established.

In the last years, particularly studies on adoption became of more and more interest in the realm of agricultural research. The ISI Web of Science counts more than 550 hits for the search string '*adoption* AND *agriculture*' for the last decade and some 120 hits for '*adoption* AND *agroforestry*' (ISI (2006)). The motivation for this particular kind of research with regard to agroforestry must, amongst other things, be seen in the irregular success rates of agroforestry projects especially in the developing world (NAIR (1996)). Promoted as favorable low external-input technologies for poor rural areas, the adoption results were often disillusioning (MOSER and BARRETT (2003)).

Slow and ineffective uptake as well as high drop out rates finally drew attention to the necessity to develop and test empirical models that predict the decision-making of farm households regarding land-use (SANCHEZ (1995)). In addition, MERCER and MILLER (1998) in their review on socioeconomic research related to agroforestry point out that adoption, i.e., the decision-making process of the household, is one of the most crucial gaps of agroforestry-related socioeconomic knowledge. Both, biophysical models that aim to predict how trees and crops interact in agroforestry systems, and purely economic techniques like cost-benefit and profitability analyses somehow 'dominated' over research approaches focusing more on socioeconomic issues. Besides that, MERCER and MILLER (1998) highlight that *"most of the economic analyses have applied rather simplistic analytical techniques to questionable data."* The majority of these studies applies dichotomous choice models (e.g., logit / probit models) to explain adoption decisions based on farm and household characteristics but do not relate the empirical analyses to an underlying theoretical model (MERCER and PATTANAYAK (2003)).

In addition, it is criticized that models based on agricultural household surveys represent only a snapshot in time (PATTANAYAK et al. (2003)). This is essentially true for all cross-sectional data and often goes back to time and budgetary constraints in the research design. However, instead of deliberately extending a cross-sectional data set via longer recall periods in the survey to a longitudinal setting, which is in essence questionable (see e.g., UCHIDA et al. (2007)), the study at hand sticks to conservativeness with regard to field research data. GROOTAERT (1982) recommends recall periods of at most one year in order to guarantee a certain accuracy of the responses. Others suggest even shorter recall periods of only a few months because the economic environment of peasant farmers is often strongly seasonal which of course influences the truthfulness of responses given during field work (CAVENDISH (2002)).

Taking into account the above-mentioned criticism with regard to the analytical techniques and the empirical modeling, a more comprehensive approach to model the farmer's decision-making process with regard to forest-related carbon finance projects has been chosen in the following. To do so, at first factor categories supposed to determine adoption are defined on the basis of the theoretical model presented in chapter 4.2. These factor categories are less narrow in their explanatory power than single descriptive variables and can simultaneously be regressed on each other using the techniques of structural equation modeling explained further down in chapter 5.

In the theoretical model above it was shown that a series of factors determine the choice variables of the household that decide upon adoption. These are risk and uncertainty (*R*) influencing the decision of the household as well as the household's resource endowments (*E*). Furthermore, biophysical farm characteristics (*Z*), the socio-demographics of the household (*H*), and market-based incentives (*I*) were critical for the household's decision. In addition, social interactions (*G*) of the participating farmers were shown to be relevant. These conceptual factor categories are 'filled with life' by describing them through suitable proxies. Chapter 4.1 portrayed such proxies by outlining the key economic and social variables for adoption cited in the literature. Based on the meta-analysis of agroforestry adoption studies from PATTANAYAK et al. (2003) the following paragraphs will subsume the various variables of adoption (i.e., economic and social ones) under one of the above mentioned factors. The aim is to group single variables to unfold more explanatory and predictive power as they are classified into factors.

The risky and uncertain environment under which a farmer must earn a livelihood is characterized by short- and long-term risks and uncertainties (*R*). All issues with regard to insecure tenure of landholdings would be an example for the latter form of risk. More short-term risks are for example fluctuations in projected output or commodity prices as well as sketchy rainfall patterns. In general, all unknowns of the market and the institutional environment influencing the decision of the farmer are described by variables grouped under the factor *R*.

The social interactions (*G*) of the participating farmers can have a potentially risk-mitigating effect as discussed above. Positive and encouraging as well as disappointing experiences and precautions regarding the new technology are communicated not only through official channels like extension services or training programs but also through community or church group memberships. The more 'connected' to those information sources a farmer is the more capable to adapt to potential risks he or she is. And, the lower the risk and uncertainty the more likely is the adoption of a new technology. The issue was discussed above under the social aspect of the adoption decision. All variables measuring information dissemination, be it via the media, or via extension services or even farmer-to-farmer contacts can be grouped under the factor for social interactions (*G*).

Chapter 4.1 discussed that assets describing the resource base of the farmer were important economic variables. Most importantly land holdings as well as other durables owned by the farmer were mentioned. Additionally, different vehicles and livestock which are often status symbols representing prosperity are of interest. Within the factor-framework these variables are subsumed under the household's resource endowments (*E*). In general, the resource endowment factor measures resources the farmer requires directly or indirectly to put the new technology into operation (MERCER and PATTANAYAK (2003)).

The socio-demographics of the household (*H*) are supposed to determine the preferences of the household. In general, preferences are difficult to measure directly. Hence, suitable proxies are required to depict the household's preference scheme. Most often, the three main socio-demographic variables, education, age, and gender are utilized. However, PATTANAYAK et al. (2003) emphasize that these proxies are debatable. The educational status for example might rather appraise the opportunity costs of labor investments in the new technology to be adopted. Likewise might the gender variable rather proxy for the labor availability of the household. As will be shown below, the structural equation modeling approach takes this issue into account by allowing complementarity (i.e., correlation) of variables grouped under different factors. Thus, the grouping proposed here is not static as will be clarified once the modeling approach is more precisely described in the following chapter 5.

The biophysical farm characteristics (*Z*) are described by soil quality and slope of the agricultural and forestry plots as well as the plot size. These variables are the crucial resources of the farm's physical production process. Implicitly, these variables are also economic determinants as they affect the production's costs and returns (MERCER and PATTANAYAK (2003)). As shown above, the correlations of the variables subsumed under this factor can be ambiguous because on the one hand poorer biophysical characteristics are supposed to positively impact on adoption. However, on the other hand some sites might be perceived by the farmer under a certain threshold justifying a new investment (PATTANAYAK et al. (2003)).

Finally, market-based incentives (*I*) were discussed to be critical for the household's decision process. Under this factor, all variables measuring either cost reductions or the generation of higher benefits can be subsumed. All typical economic determinants of adoption fall in this category, like variables measuring income from agricultural activities, off-farm income, and credit availability. As markets are thin in subsistence economies and data material is often limited (e.g., cross-sectional data), sometimes even proxies for prices must be established.

As mentioned before, the grouping of variables proposed here is not static but a first starting point for the structural equation modeling developed in the following chapter.

5. Structural Equation Modeling (SEM) to represent farmer decisions

In this chapter, a structural equation model (SEM) is specified, estimated, tested and modified. Subsequently, the results are presented and discussed in detail. Referring to modification, the SEM approach, taken here, includes a feedback mechanism to eventually improve the model. The basic approach of SEM is a *confirmatory* one, i.e., hypotheses are put forward and subsequently confirmed or rejected by the model without modifying the model. However, JÖRESKOG (1993) introduced a more concise differentiation between three different strategies for structural equation modeling. A *strictly confirmatory* strategy, the strategy of *alternative models*, and a *model generating approach*. Following the first strategy, in the purely confirmatory mode, the researcher hypothesizes one single model based on a theory.

Subsequently, sample data is collected and the fit of the hypothesized model to the sample data is tested. Based on this test the researcher either accepts or rejects the model. It is important to note that no modifications of the model are foreseen in this strictly confirmatory mode. By choosing the strategy of alternative models, the researcher a priori establishes several different models. As a matter of course, all of these have to be based on sound theory underlying the SEM approach. Based on the data analysis and the model fit, the researcher then selects the model best representing the sample data. In the third case, the researcher postulates an initial model that is modified in case of its rejection.

As will be shown, the modification process has the goal to (i) locate the source of model misfit and (ii) to respecify the model accordingly. Although the strictly confirmatory approach of SEM is abandoned once the model is respecified, the third approach is the one most commonly applied by researchers (BYRNE (2001)). The underlying intention of the modification is to find a substantively meaningful and statistically well-fitting model in the end. The argumentation for the respecification is justifiable given the costs associated with field research and questionnaire-based data collection. The following chapters therefore follow a specific procedure to develop and analyze a structural equation model depicting the farmer's participation decision. Fig. 5-1 sketches the consecutive steps of this procedure.

A concise problem formulation is the starting point for any SEM analysis. Chapter 4 introduced the problem setting by assessing variables determining farmers' participation in new agricultural technologies and carbon finance projects. This was done on the basis of the above literature review with regard to agroforestry adoption. Chapter 5 concretizes the problem by applying the theory of chapter 4 to an empirical data set. The SEM analysis is a necessary first step in the study of potential project impacts. In SEM modeling, the establishment of hypotheses prior to the modeling is usually split into the formulation of *structural* and *measurement hypotheses* (see Fig. 5-1). In this step, the initial hypotheses of chapter 1.1, which were subsequently discussed in chapter 4.3, are elaborated in more detail and fine-tuned to fit into the specific SEM framework. Furthermore,

Fig. 5-1 shows that empirical data collection (i.e., the field survey) follows next. Then, the stepwise approach for SEM analysis proceeds with the specification of a SEM, the estimation, and the testing of the SEM. A feedback loop allowing for model modification and re-specification is shown on the left hand side of Fig. 5-1 before model results are finally interpreted.

Fig. 5-1 Stepwise Approach for the Analysis of Structural Equation Models

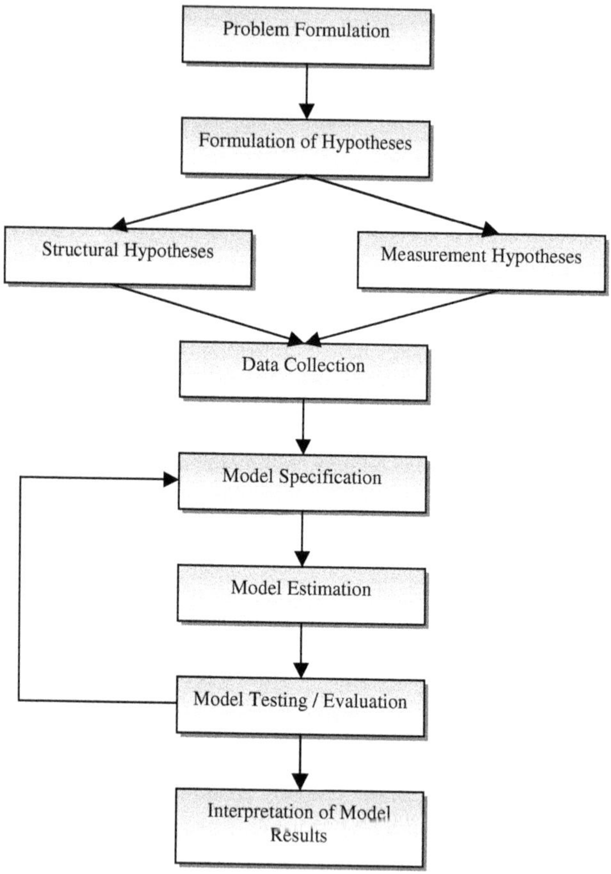

Source: REINECKE (2005) - modified

At first, chapter 5.1 will briefly introduce the general concept of structural equation modeling and describe in a few words the software package used for modeling purposes.

5.1. The Concept of Structural Equation Modeling

In a structural equation model different types of models, i.e., regression, path, and factor analytic models are interacting simultaneously to analyze a specific problem under consideration. As outlined above, structural equation models generally take a confirmatory (i.e., hypothesis-testing) approach to problem analysis. The processes under study are usually represented by a series of structural (i.e., regression) equations. The hypothesized model is tested statistically for its consistency by simultaneous analysis. While in more common statistical applications like multiple regression or ANOVA (analysis of variance) the *individual* observations are modeled, SEM procedures feature *covariances* rather than cases. Or, as BOLLEN (1989) puts it: *"Instead of minimizing functions of observed and predicted individual values, we minimize the difference between the sample covariances and the covariances predicted by the model.*"

The fundamental hypothesis underlying all structural equation models is that the covariance matrix of the observed variables in essence is a function of a set of parameters. If the established model were correct and its parameters known, the population covariance matrix would be exactly reproduced. More formalized the above mentioned concept can be expressed as:

$$\Sigma = \Sigma(\theta) \tag{5.1}$$

where Σ is the population covariance matrix of the observed variables, θ is a vector that contains the model parameters, and $\Sigma(\theta)$ is the covariance matrix written as a function of θ.

Hence, $\Sigma(\theta)$ stands for the restricted covariance matrix as specified by the hypothesized model. Equation (5.1) is the elementary null hypothesis (H_0) being tested in SEM. However, in contrast to traditional statistical procedures the researcher hopes *not* to reject H_0, i.e., it is desired that the postulated model is true for the population (BYRNE (2001)). Another difference between traditional multivariate procedures and SEM is the fact that SEM can take potential measurement error into account. While the former assume that error in the explanatory variables vanishes, SEM makes explicit estimates of these error variance parameters. Thus, the rather strong assumptions of techniques rooted in regression can eventually lead to inaccuracies which can be taken into consideration by SEM (HOYLE (1995)). Furthermore, data analyses with traditional multivariate procedures are based on directly measured (i.e., observed) variables, while SEM procedures can incorporate both, observed variables as well as *unobserved* variables. These unobserved variables are not directly measurable and therefore called '*latent variables*' or factors. The concept of latent variables originated from the behavioral sciences where researchers have a specific interest in studying theoretical constructs that can not be observed directly, e.g., motivation (LOEHLIN (2004)). An economic example for a latent variable would be 'social class' which per se cannot be directly determined but measured by a set of other variables termed *observed* or *manifest* variables. It is important to emphasize that latent variables are not directly observed and hence cannot be directly measured. This implies that the researcher has to operationally define the latent variables by suitable indicators.[46] Within SEM, latent variables can become both, *exogenous* and *endogenous*. An exogenous latent variable is an independent variable. Endogenous latent variables are dependent variables that can either be directly or indirectly influenced by the exogenous ones. As all latent variables that potentially influence an endogenous variable are included in a correctly specified SEM, the fluctuation of the endogenous variables is assumed to be completely described by the model.

To get back to the initial statement that SEMs take a confirmatory approach to problem analysis, the different elements of a full-fledged SEM are briefly outlined in the following. A complete SEM comprises two models: a *measurement* as well as a *structural model*. The measurement model describes the relation between the latent variables and their measured indicator variables while the core structural model depicts the impact of one latent construct (i.e., variable) on another. The measurement model makes use of a very well established statistical procedure, namely factor analysis - but in a *confirmatory* mode.

Factor analysis in general uses the correlation coefficient to determine observed variables creating a latent variable or factor.[47] While traditionally factor analysis is used as a tool for data reduction in a rather exploratory mode, it is important to note that in SEM the researcher establishes the measurement model, i.e., the confirmatory factor model by hypotheses. This is done based on prior knowledge of the subject matter, e.g., theory, body of literature, or field research. The researcher postulates a priori relationships between the observed and the latent variables, which are then tested statistically. In summary, the measurement model consists of a confirmatory factor analytic model that explains the extent to which the observed variables are generated by the underlying constructs (BYRNE (2001)). The regression paths from the factors to the observed variables (i.e., the so-called factor loadings) are of primary interest in the measurement model. As mentioned earlier, the fundamental structural model then depicts the relationships between latent constructs, i.e., inter-factor relations or in other words the regression structure among the latent variables.

While in the past SEMs could be solely represented and modeled by a series of structural equations, the new feature of the software program AMOS®, used for this study, allows the structural relations underlying a SEM to be modeled pictorially.[48] AMOS® has two different modes of operation called 'AMOS® Basic' and 'AMOS® Graphics'. While the basic mode of AMOS® is a programming interface where models can be specified via equations, AMOS® graphical mode will be

46 The terms *latent*, *unobserved* and *unmeasured* are often used interchangeably to describe a hypothetical construct, i.e., a factor or latent variable. The terms *observed*, *manifest*, and *measured* variable are also used synonymously.

47 The terms latent variable and factor have the same meaning in this context.

48 AMOS® is an acronym that stands for 'Analysis of Moment Structures', i.e., analysis of covariance structures.

mainly discussed here. It allows the researcher to specify SEMs with the help of a graphical interface. SEMs can be formulated by different symbols at first and subsequently linked to the underlying data set for analysis. The visualization of the SEM in AMOS® Graphics allows a clear conceptualization of the theory and hypotheses under study (ARBUCKLE and WOTHKE (1999)).

Per convention, SEMs are depicted by using the following symbols: squares (rectangles) represent observed and directly measured variables, ellipses (circles) represent unobserved variables of the model, single-headed arrows (→) represent the impact of one variable on another, while double-headed arrows (↔) stand for covariances or correlations between two variables. These symbols put together to form a schematic model representation are termed a *path diagram*. Such path diagrams or path models aimed to portray assumed relationships of the variables under study were first developed by WRIGHT (1918); (1921), a biometrician who dealt with models of animal behavior. Today, SEM is a multidisciplinary modeling approach used not only in biology but also in sociology, psychology and in economics. In the realm of economics GOLDBERGER (1972) is considered to be one of the 'pioneers' of structural equation modeling.

In summary, a path diagram of a particular SEM is the graphical equivalent of the underlying mathematical SEM representation which is in essence a set of simultaneous equations as will be seen further down. Fig. 5-2 illustrates a basic, simplified structural equation model graphically. Subsequently, the model shown is explained in more detail below.

Fig. 5-2 Basic Structural Equation Model

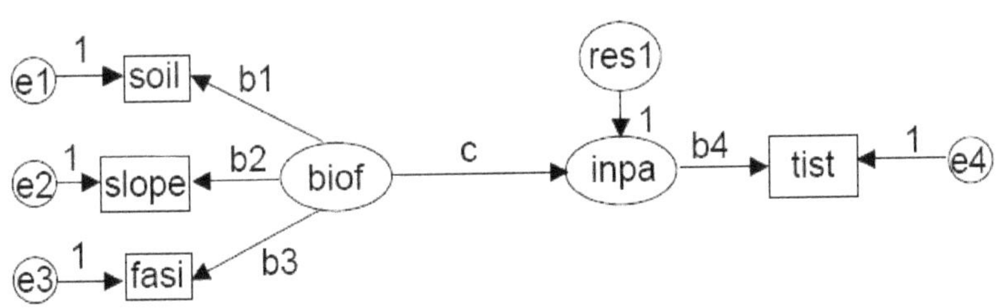

In the center of the model, a hypothesized impact *c* of the variable *biof* (biophysical farm characteristics) on the inclination of farmers to participate (*inpa*) in the TIST program shall be observed. A single-headed arrow (→) graphically represents the theoretically assumed impact. The variables *biof* as well as *inpa* are both unobserved, latent variables (graphically represented as ellipses) which are measured by their respective indicators (the rectangles). The bio-physical characteristics of the farm are measured by properties of the *soil*, the *slope* and the size of the farm (*fasi*). The inclination to participate in the program (*inpa*) is measured here by actual participation via the dichotomous *tist* variable only. Each observed variable carries an error term (*e1* – *e4*) and the predicted inclination to participate (*inpa*) carries a residual error term (*res1*).[49] Symbols representing variables are thus linked by arrows that symbolize hypothesized processes amongst those variables. The unidirectional arrows from the latent variables *biof* and *inpa* to their respective indicators stand for structural or path coefficients representing the magnitude of expected change in the manifest variables for every change in the related factor (i.e., factor loading). The single-headed arrows pointing from the encircled error terms to the measured variables visualize measurement error, while the arrow from *res1* to *inpa* symbolizes the impact of error in the prediction of the farmers inclination to participation based on the farm's bio-physical factors. In this simple model, no double-headed arrows are included. Such arrows could be used to represent correlation between two variables.

For the sake of the concept's completeness the transformation of the path diagram into structural equations is clarified in the following. As pointed out earlier, arrows symbolize influences from one variable on another as in regression equations where the influence of the predictor(s) on the dependent variable is formulated. Thus, by formulating the structural equations every variable that has one or more arrows pointing towards it is noted down as dependent variable and all influencing variables are part of the respective equation's right hand side.[50] In the example above five variables have arrows pointing towards them indicating five simultaneous equations for five endogenous variables. Equation (5.2) represents the core structural model linking the two latent variables *biof* and *inpa*.

$$inpa = c * biof + res_1 \quad (5.2)$$
$$soil = b1 * biof + e_1 \quad (5.3)$$
$$slope = b2 * biof + e_2 \quad (5.4)$$
$$fasi = b3 * biof + e_3 \quad (5.5)$$
$$tist = b4 * inpa + e_4 \quad (5.6)$$

The other equations establish two measurement models being part of the basic structural equation model shown in Fig. 5-2, namely one measurement model

49 Measurement error is associated with an observed (measured) variable and reflects the (in)adequacy in measuring the indicators. In principle, this measurement error could be split into a random and a nonrandom or systematic error. Residual error terms are inherent to the prediction of a latent variable (i.e., an endogenous factor) by exogenous factors. According to the representational rule of unobserved variables in path modeling errors are depicted as circles or ellipses.

50 It is important to note here that the graphical representation of a SEM works from left to right with the exogenous latent variables usually on the model's left hand side and the endogenous latent variables on the right. This might somehow contradict the mathematical representation at first view.

for *biof* with equations (5.3), (5.4), and (5.5) and another one for *inpa* consisting in this case only of equation (5.6).[51]

The discussion above shows that there is a one-to-one equivalence between the graphical representation of a SEM and its underlying structural equations. However, the variance of the exogenous variables that is essential for model estimation is not depicted in the two SEM representations. AMOS® by default estimates the variances of all independent variables and therefore assists the researcher in specifying a SEM. In addition, it is important to note that AMOS® does not assume any relationship between variables that are not intentionally connected by arrows (ARBUCKLE and WOTHKE (1999)).

Going once again back to Fig. 5-2, it can be seen that arrows in each case only go from one variable to another and not vice versa. Such a model is called a *recursive model* while a model allowing for reciprocal or feedback effects, i.e., arrows going from one variable to another and back at the same time, is termed non-recursive model.

In order to come to well-defined results SCHUMACKER and LOMAX (2004) in accordance with Fig. 5-1 postulated that all SEM analyses should follow a logical sequence of five steps, namely, a) model specification, b) model identification, c) model estimation, d) model testing, and e) model modification. The next paragraphs in this chapter will follow this sequence beginning with 5.5.2 on model specification and identification.

5.2. Model Specification and Identification

During model specification, relevant theory, other research findings, and additional information like the previous field research experience were brought together to develop a theoretical model. The synopsis of the abovementioned information sources guides the decision which variables to include in the hypothetical model and how these are related. In theory, a model is properly specified if the true population model is considered consistent with the theoretical model being tested. For any SEM application, the sample covariance matrix has to be 'sufficiently' reproduced by the implied theoretical model.[52] In other words, the research goal is to specify the best possible model that reproduces the sample covariance matrix (SCHUMACKER and LOMAX (2004)). Finally, this 'true model' meaning the model that generated the data has to be compared to the implied theoretical model. If the true model deviates from the implied theoretical model, the theoretical model might be mis-specified. Both, omission and/or unnecessary inclusion of variables can lead to misspecification problems. The consequence of misspecification is biased parameter estimates and the theoretical

51 The ones (1) assigned to the arrows going form the error terms to the respective variables indicate a fixed value of one, which means that these regression path parameters are not estimated.

52 Chapter 5.4. on model testing will discuss this issue further.

model results may not fit the actual data well. The subsequent chapters dealing with model estimation, testing and modification will further elaborate on the issue of model misspecification and its detection.

The initial model presented below is built on the analysis of chapter 4. The factor categories hypothesized to explain adoption were identified in the theoretical model of chapter 4.2 based on the work of PATTANAYAK et al. (2003). These factor categories are now combined to construct an initial structural equation model. In chapter 4.2 it was shown that several factors are supposed to determine the household's decision with regard to adoption: risk and uncertainty (*R*), the household's resource endowments (*E*), biophysical farm characteristics (*Z*), the socio-demographics of the household (*H*), and market-based incentives (*I*). In addition, social interactions (*G*) of the participating farmers played a relevant role. In a first step with regard to structural equation modeling, suitable proxies concretized these conceptual factor categories. The aim was to increase the explanatory power by subsuming single variables under the factors. The initially specified model is graphically represented with AMOS® in Fig. 5-3.

The left hand side of the SEM shown in Fig. 5-3 displays six different measurement models.[53] The six latent variables (ellipses) on the left are hypothesized to influence the household's inclination to participate, *inpa*, on the right hand side of the model. For reasons of simplicity the factors are slightly renamed in the following. The socio-demographics of the household (*H*) are operationalized by the latent variable '*preferences*' (*pref*). It is assumed to best measure those preferences by the educational status of the household head (*educ*), the age of the household head (*age*) and the household head's gender (*gndr*). It is important to note that all the path coefficients going from the latent variable *pref* to their respective indicators (*educ*, *age*, *gndr*) are measurement hypotheses within the model. This is equally true for all the other path coefficients of the six measurement models on the left hand side of Fig. 5-3.

The resource endowments (*E*) of the household are depicted by the latent variable *rend* and measured by the observed variables explained in the following.[54] First, the sum of the household's general expenditures (*s_exp*) du-ring the last twelve months as a proxy for income.[55] This includes all household expenditures except for food, e.g. medicine and health costs, expenditures for clothing and shoes, festivities, transport and communication, rent of agricultural equipment or land, and expenditures for eventually external hired labor.

53 The measurement models are basically sub-modules of the structural equation model.

54 Again, these are all measurement hypotheses as other unknown proxies might be equally suitable to depict *rend*.

55 The household's general expenditures during the last year were taken as a proxy for income in general as it is often very sensitive to directly ask farmers for their actual gross income. For a more detailed discussion of this proxy see chapter 3.2.2 and MORRIS et al. (2000); DEATON and GROSH (2000).

Fig. 5-3 Initial Structural Equation Model

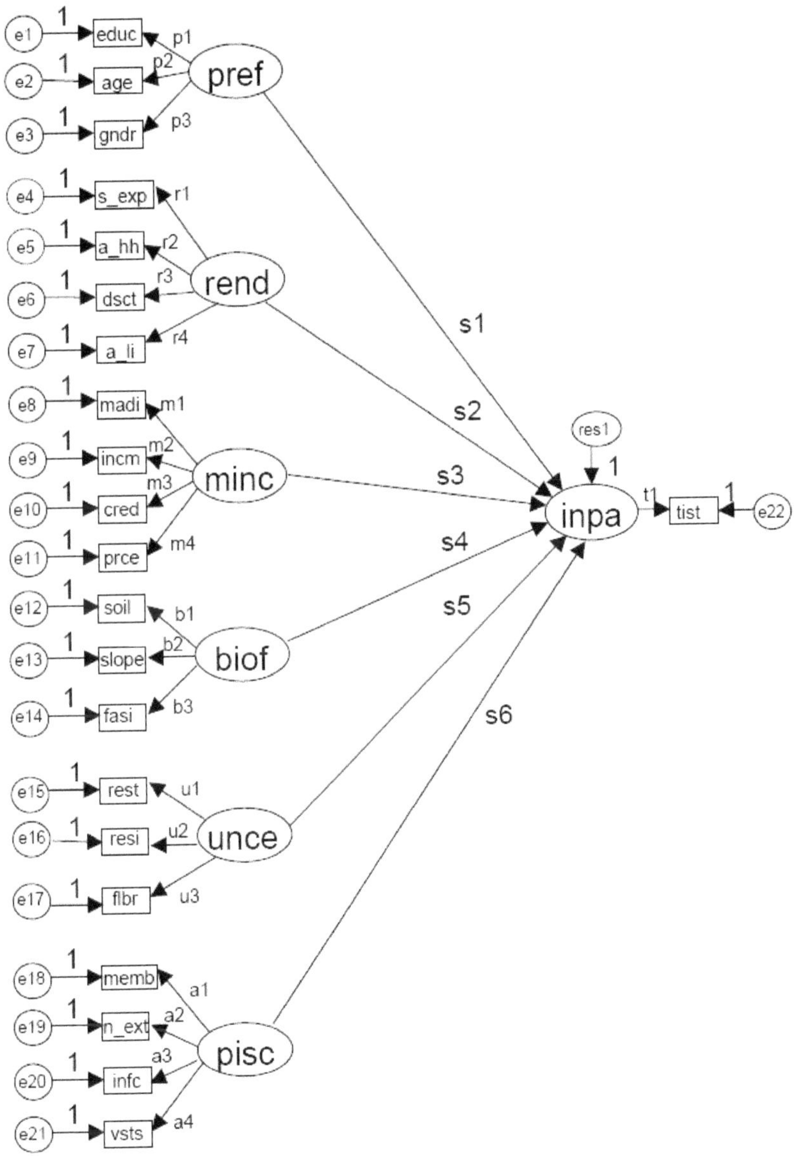

In addition, other farm related expenditures for seeds, tools etc. are included. Furthermore, other household consumables, e.g. soap, kerosene etc. and costs for education (i.e., school fees, uniforms, books etc.) are comprised. Finally, expenses like assistance to relatives or friends and costs for funerals are accounted for. Moreover, household assets (*a_hh*) are an observed variable, which is hypothesized to add to the overall resource endowments of the household. Household assets translate different kinds of household durables into monetary terms (e.g., vehicles such as bicycles, motorbikes, eventually tractor or car etc.). The same monetary assumption is made for other assets necessary for agricultural production (e.g., hoes, axes, rakes, animal pulled plow, ox cart etc.). Furthermore, assets determining a certain standard of living in rural Tanzania, e.g. radio, chair, table, bed, mattress are taken into account.[56] Similarly, the variable measuring livestock ownership of the household (*a_li*) converts the livestock of the household into monetary value measured at local market prices. The fourth measured variable under the resource endowments (*rend*) of the household is a proxy for the labor availability within the household by measuring household members, or descendants (*dsct*), able to assist in farming activities besides the household head.

The market-based incentives (*I*) of the theoretical model in chapter 4.2 are represented by the latent variable *minc* which is also measured by four manifest variables. The distance to the next marketplace to buy and sell agricultural goods (*madi*), the off-farm income (*incm*) a household can realize through other business activities except from farming or tree planting activities (e.g., burning of bricks, repairing of bicycles etc.), the availability of credit to start a new agricultural business (*cred*), and the price for capital (*prce*) as a general price proxy.

The biophysical farm characteristics (*Z*) become the latent variable *biof* in the structural equation model. According to MOWO et al. (2006) self-reported soil (*soil*) and slope (*slope*) properties of the plots are hypothesized to measure the biophysical attributes of the farm adequately.[57] In addition, the size of all plots for agricultural production or tree planting is measured by the farm size (*fasi*).

The factor category of risk and uncertainty (*R*) is represented in the structural equation model by the latent variable *unce*, which is measured by three manifest variables (*rest*, *resi* and *flbr*). The absence of tenure security through legal title or other ownership rights like customary laws, restrict the usage of certain plots the farmer owns. The variable *rest* measures the size of all plots facing any form of usage restriction and proportionally relates it to the whole farm size. In addition, two other manifest variables are assumed to proxy for the

56 Especially household assets like a mattress turned out to be sensitive parts of the survey questionnaire. Rural households often own traditionally manufactured beds with a customary form of mattress consisting of special sisal netting. An industrially produced mattress is considered as a form of luxury good in rural areas (ZABLON (2005)).

57 For a more detailed discussion of those self-reported values see chapter 3.3.4.

risky environment a farmer faces. One is the years of residence (*resi*) of the particular household in the community with more risky environments making the household move more often in order to earn a livelihood elsewhere, while the other manifest variable describes the overall labor availability in the extended family (*flbr*) to counteract seasonal labor shortages. While PATTANAYAK et al. (2003) subsume access to extension and training (*n_ext*) and household membership in different community organizations (*memb*) under the factor category of risk and uncertainty, the initial structural equation model presented in Fig. 5-3 puts forward an alternative differentiation.

A new latent variable envisaging the aforementioned social interactions (G) of the participating farmers is constructed. It describes a perceived indicator of social capital (*pisc*) which is measured amongst others by the above-named variables *memb* and *n_ext*.[58] Additionally, the variable *infc* assesses all the sources for agricultural market and price information used by the farm household like newspaper(s), radio, conversation with extension staff or farmer leaders, conversation with friends and/or relatives in cooperatives, associations, church groups etc. Finally, the variable *vsts* delivers additional information on the connectedness of the farm household by counting all the actual visits of different meetings in community or church groups, trade unions, cooperatives or associations other than the Small Group meetings of the TIST program.

All indicator variables of the six measurement models on the left are influenced by measurement error, as indicated by the associated error terms (*e1 – e21*). Each of the 21 indicator variables is regressed onto its respective factor, *pref*, *rend*, *minc*, *biof*, *unce*, or *pisc*. Subsequently, the six latent variables are regressed on the household's inclination to participate, *inpa* (right hand side of the model) as depicted by the straight single-headed arrows (→) going from the left hand side of the model to the right. The arrows represent six structural models and hence six structural hypotheses depicting the relationships between the latent variables on the left and the latent variable on the right hand side of the model. These inter-factor relations or regression structures among the latent variables are at the center of the initial structural equation model in Fig. 5-3. As in the basic structural equation model of Fig. 5-2 the inclination to participate in the program (*inpa*) is measured by actual program participation by the dichotomous *tist* variable, which is influenced by its associated error term (*e22*). Furthermore, *inpa* carries a residual term (*res1*) measuring inherent error to the prediction of one endogenous latent variable by exogenous factors, i.e., the six latent variables on the left hand side of the model.

All information necessary for model specification is summarized in Table 5-1 below. The name of the variables and their respective label (columns two

---

58 The perceived indicator of social capital measures the '*connectedness*' of the respective farm household within the community based on self-reported data. While NARAYAN and PRITCHETT (1999) for example construct a 'social capital index' the perceptional aspect of self-reported data should be noted here.

and three) as well as the type of the variables within the structural equation model, i.e., latent exogenous (L-EX), latent endogenous (L-EN), or measured endogenous (M-EN) are shown (column one). In addition, column four indicates the measurement level of the observed variables. Since latent variables are not directly measured, they initially have no explicit scale determined. Further down it will be seen that constraining one of the observed variables loading on a latent one will determine the scale of the construct.

Table 5-1 Description of Variables

Variable Type	Variable Name	Variable Label	Measured as / in	Measurement / Structural Hypotheses
L-EX	Preferences	pref	(+) Household preferences characterize the objectives and motivations deciding upon participation.	
M-EN	Education	educ	education level of household head (years of schooling)	(+) New technology demands more skills
M-EN	Age	age	age of household head in years	(+) age of farmer linked to longer farming experience and greater accumulation of knowledge; (-) shorter planning horizons and higher risk aversion of older farmers
M-EN	Gender	gndr	proportion of male household members	(+) higher male proportion more likely to adopt new technology
L-EX	Resource Endowments	rend	(+) Enable the technology choice of the household.	
M-EN	Expenditures	s_exp	sum of expenditures per year in Tanzanian Shillings taken as a proxy for income	(+) correlation of income with adoption is typically positive
M-EN	Assets	a_hh	value of durables in Tanzanian Shillings	(+) greater wealth makes the investment in a new technology more feasible
M-EN	Descendants	dsct	descendants in the household working in the agricultural production	(+) greater labor availability promotes adoption
M-EN	Livestock	a_li	value of livestock in Tanzanian Shillings	(+) greater wealth makes the investment in a new technology more feasible
L-EX	Market Incentives	minc	(-) Focus on the explicit economic determinants of potential adoption	
M-EN	Market distance	madi	kilometers to walk	(-) greater market distance impedes adoption
M-EN	Off-farm income	incm	off-farm income in Tanzanian Shillings	(-) farmers mainly relying on off-farm income might have no interest to adopt a new technology

M-EN	Credit /Savings	cred	credit availability (binary)	(+) credit availability has positive correlation with adoption
M-EN	Price	prce	capital price variable (8-item scale, ordinal)	(-) higher price for capital higher costs of the investment in the new technology
L-EX	Bio-Physical Factors	biof	(+) Influence the physical production process	
M-EN	Soil	soil	self-reported soil quality (5-item scale, ordinal)	(-) superior soil quality is negatively correlated with adoption
M-EN	Slope	slope	self-reported slope (6-item scale, ordinal)	(+) the steeper the plots, the more likely to adopt
M-EN	Plot Size	fasi	acres	(+) greater holdings make the investment in a new technology more feasible
L-EX	Risk and Uncertainty	unce	(-) Unknowns in the market and institutional environment	
M-EN	Tenure	rest	proportion of plots with restricted access (proportional tenure restriction measure)	(-) landowners are more likely to adopt a new technology than tenants
M-EN	Residence	resi	years	(-) riskier environments making the household move more often
M-EN	Labor	flbr	size of the extended family	(+) size of the family determines labor availability
L-EX	Social Capital	pisc	(+) Perception-based index of social capital, measures connectedness	
M-EN	Membership	memb	count of memberships in different organizations	(+) greater social activity promotes knowledge about the new technology
M-EN	Extension	n_ext	count extension contacts in the last 12 months (count)	(+) extension contacts usually positively related with adoption decision
M-EN	Market & Price Information	infc	count of sources of information	(+) information channels promote knowledge about the new technology
M-EN	Visits	vsts	count visits per year in organizations	(+) greater social activity promotes knowledge about the new technology
L-EN	Inclination to participate	inpa	Latent construct measuring farmer's participation inclination	
M-EN	TIST membership	tist	participation (binary)	actual participation decision by the respective farm households

The last column summarizes the measurement hypotheses (for the observed variables), or structural hypotheses for the relationships amongst the latent variables, and additionally hypothesizes an algebraic sign for the respective relationship.

Following the rules described in chapter 5.1 for the transformation of a path diagram into the underlying structural equations, the initial structural model of Fig. 5-3 is represented mathematically in equations 5.6 through 5.28.

Measurement model 1:
latent *exogenous* variable *pref*

$$educ = p_1 * pref + e_1 \quad (5.6)$$
$$age = p_2 * pref + e_2 \quad (5.7)$$
$$gndr = p_3 * pref + e_3 \quad (5.8)$$

Measurement model 2:
latent *exogenous* variable *rend*

$$s\_exp = r_1 * rend + e_4 \quad (5.9)$$
$$a\_hh = r_2 * rend + e_5 \quad (5.10)$$
$$dsct = r_3 * rend + e_6 \quad (5.11)$$
$$a\_li = r_4 * rend + e_7 \quad (5.12)$$

Measurement model 3:
latent *exogenous* variable *minc*

$$madi = m_1 * minc + e_8 \quad (5.13)$$
$$incm = m_2 * minc + e_9 \quad (5.14)$$
$$cred = m_3 * minc + e_{10} \quad (5.15)$$
$$prce = m_4 * minc + e_{11} \quad (5.16)$$

Measurement model 4:
latent *exogenous* variable *biof*

$$soil = b_1 * biof + e_{12} \quad (5.17)$$
$$slope = b_2 * biof + e_{13} \quad (5.18)$$
$$fasi = b_3 * biof + e_{14} \quad (5.19)$$

Measurement model 5:
latent *exogenous* variable *unce*

$$rest = u_1 * unce + e_{15} \quad (5.20)$$
$$resi = u_2 * unce + e_{16} \quad (5.21)$$
$$flbr = u_3 * unce + e_{17} \quad (5.22)$$

Measurement model 6:
latent *exogenous* variable *pisc*

$$memb = a_1 * pisc + e_{18} \quad (5.23)$$
$$n\_ext = a_2 * pisc + e_{19} \quad (5.24)$$
$$infc = a_3 * pisc + e_{20} \quad (5.25)$$
$$vsts = a_4 * pisc + e_{21} \quad (5.26)$$

Measurement model:
latent *endogenous* variable *inpa*

$$tist = t_1 * inpa + e_{22} \quad (5.27)$$

Structural model:

$$inpa = s_1 * pref + s_2 * rend + s_3 * minc + s_4 * biof + s_5 * unce + s_6 * pisc + res_1 \quad (5.28)$$

The measurement models summarize the influence of the respective latent variables on the observed variables given by the factor loadings or path coefficients (p_n, r_n, m_n, b_n, u_n, a_n and t_n). In addition, the error terms (e_1-e_{22}) influence the observed variables. Generally, only the variance of the measurement error is of interest in the model. This is expressed by fixing all regression paths from the error terms to the observed variables to 1. In order not to detract from clarity, the equation system outlined above follows the assumptions underlying model estimation with the AMOS® Basic programming interface, namely that all exogenous factors can be inter-correlated and the error terms are uncorrelated (ARBUCKLE and WOTHKE (1999)).

After specifying the initial model and describing its major features, the issue of *model identification* has to be addressed next. The specific question with regard to model identification is the following: Is it possible to find a unique set of parameter estimates that is consistent with the sample data (contained in the sample covariance matrix) and the theoretical model (implied by the population covariance matrix)? If the data fits more than one model equally well there is so called *indeterminacy* (SCHUMACKER and LOMAX (2004)). If on the contrary a unique solution for the values of the structural parameters can be found, the model is identified. As a consequence the parameters are estimable and the model therefore testable. In SEM each model parameter is either of the following: a *free* parameter that is unknown and therefore to be estimated or a parameter *fixed* to a pre-specified value (see the path coefficients of the error terms above which are fixed to 1), or a *constrained* parameter set equal to one or more other parameters. Model identification hinges on the actual type of parameters as being fixed, free or constrained. In general, SEMs may be just-identified, overidentified, or underidentified. If a SEM is just-identified there is a one-to-one correspondence between the data and the structural parameters. A just-identified model is of minor scientific interest as it has no degrees of freedom and can never be rejected. An overidentified model is one in which the number of data points (i.e., the variances and covariances of the observed variables) is greater than the number of parameters to be estimated. This situation results in positive degrees of freedom that allow for potential rejection of the model – a fact that makes the model 'useful' from a scientific point of view.

The actual aim in SEM is to specify a model that meets the criterion of overidentification (BYRNE (2001)). Lastly, an underidentified (or not identified) model is one in which the number of data points is less than the parameters to be estimated. In this case the degrees of freedom are negative and the parameter estimates are unreliable. All current SEM software including AMOS® provides the researcher with information regarding model identification. Nevertheless, it is desirable to actually verify how many variances and covariances (i.e., data points) there are in any specific model and weigh this information against the model parameters to be estimated.

With p observed variables there are $p(p+1)/2$ data points available for parameter estimation (BOLLEN (1989)). For the initial model presented in Fig. 5-3 with 22 observed variables this means that there are 253 variances and covariances of the observed variables available for model estimation. Comparing these data points with the parameters to be estimated, one can judge whether the model discussed is over-, under-, or just-identified. The initial model of Fig. 5-3 requires the following parameters to be estimated: 22 regression parameters, i.e., the factor loadings as depicted by the single-headed arrows. In addition, 22 variances of the error terms, the variance of the residual term and seven factor variances (graphically not represented) have to be estimated. Finally, 15 factor covariances have to be calculated. With the data points outweighing the parameters, the initial model is overidentified and has positive degrees of freedom.

As BYRNE (2001) points out the specification of an overidentified model is a necessary but not sufficient condition to resolve the identification problem.[59] In the practice of modeling the imposition of constraints can often solve identification problems. These constraints are on the one hand important for model identification and on the other hand for the issue of the scale of the latent variables in the model. As mentioned above, the latent variables are not directly measured and therefore initially have no explicit scale determined. By constraining one of the measured variables loading on a latent one, typically to 1.0, the scale of the construct is determined and ideally the identification problem is resolved.[60] This means that for each latent variable there is one path coefficient to be constrained before the model presented in Fig. 5-3 can be estimated.[61]

5.3. Model Estimation

Having constrained at least one of the regression paths of each latent variable going to one of its observed variables in the measurement models above, the next step is to estimate the initial model. This section briefly introduces different methods for model parameter estimation and subsequently discusses the estimates of the initial model.

To avoid ambiguity, it is important to note that the term *model parameter* in the following means estimates for the population parameters in a SEM. The ultimate goal of the estimation procedure is to find estimates of the parameters specified in the model that produce the closest possible model implied matrix $\Sigma(\theta)$ compared to S, which is the sample covariance matrix of the observed indicator variables. In case the sample covariance matrix S is equal to the model-implied matrix $\Sigma(\theta)$, the difference of $S - \Sigma(\theta)$ would be zero. Within the estima-

---

59 There are four other different identification rules apart from the more intuitive, so called 't-Rule' described here (see BOLLEN (1989) for more details). As SCHUMACKER and LOMAX (2004) state, proofs of these conditions are often problematic in practice. To avoid going too much in detail, the reader is referred to BOLLEN (1989) or JÖRESKOG and SÖRBOM (1988) here.

60 The measured variable with the factor loading fixed is sometimes referred to as 'reference variable'. BYRNE (2001) points out that both, the decision which path to constrain as well as to which value the factor loading should be constrained, is in principle arbitrary. Usually the indicator with the highest reliability should be constrained if this information is available.

61 Alternatively, if no path coefficient is to be constrained, the variance of the latent variable must be fixed to 1.

tion process, the difference between $\Sigma(\theta)$ and S is minimized by a model-specific discrepancy function.

AMOS® provides the researcher with five different discrepancy functions for the estimation process, namely the Maximum-Likelihood method (ML), the method of Unweighted Least-Squares (ULS), the Generalized Least-Squares (GLS) method, the Scale Free Least-Squares (SLS) method, and the Asymptotically Distribution-Free (ADF) method (ARBUCKLE and WOTHKE (1999)). The actual decision which estimation method to use hinges on four criteria with regard to the input data and some desired model features (BACKHAUS et al. (2006)), namely:

- the distributional properties of the measured variables,
- the scale invariance of the fit function,
- the required sample size, and
- the availability of inferential statistics.

The most important decision criterion is the distributional properties of the observed variables. ML and GLS methods assume *normally* distributed observed variables. The other three estimation methods can even be used if the distribution of the measured indicators is not multivariate normal. The second criterion concerns the scale invariance of the estimated parameters. From the aforementioned discrepancy functions only ULS estimators are scale dependent, i.e. changes in the observed variable scales yield different estimation solutions. However, if the measured variables are standardized before the covariance matrix is computed the scale dependency can be by-passed (LONG (1983)). The third criterion regarding the choice of the estimation function is the available sample size.

In general, all SEM estimation procedures require fairly large sample sizes. Depending on the literature, sample sizes with $n \geq 100$ or $n \geq 200$ can be found (ARBUCKLE and WOTHKE (1999); LOEHLIN (2004); BENTLER (1985)). The Asymptotically Distribution-Free (ADF) method requires an even larger sample size due to the computation of the asymptotic covariance matrix (JÖRESKOG and SÖRBOM (1989)). The fourth criterion relates to the availability of inferential statistics which are necessary to assess model fit as will be seen further down. In general, the aim is to test the null hypothesis (H_0) as stated in equation (5.1), which says that the restricted covariance matrix as specified by the hypothesized model has to be equal to the population covariance matrix of the observed variables. The inferential statistic of most interest in this regard is the so-called χ^2 (chi-square) value. Chi-square is in essence a measure for the difference between the sample variance-covariance matrix and the model-implied (i.e., reproduced) variance-covariance matrix. The difference between these two matrices is contained in a residual matrix which can be requested as part of the AMOS® output. If the residual values in the residual matrix are close to zero, there is little difference between the sample variance-covariance matrix and the reproduced variance-covariance matrix. This in turn means that the chi-square value will be close to zero, indicating that the theoretically specified model fits the sample data well. Such information is only available for the ML, GLS, and ADF model estimation methods. Table 5-2 summarizes the different discrepancy functions for model estimation and the outlined selection criteria.

Table 5-2 Model Estimation: Decision Criteria for Different Discrepancy Functions

Decision Criterion	ML	GLS	ULS	SLS	ADF
multivariate normality of observed variables assumed	yes	yes	—	—	—
scale invariance	yes	yes	—	yes	yes
sample size	>100	>100	>100	>100	>200
inferential statistics	yes	yes	—	—	yes

Source: BACKHAUS et al. (2006) - modified

AMOS® by default draws on the ML estimation method which delivers the most precise parameter estimates if the aforementioned conditions with regard to sample size and distribution of the observed variables are met (ARBUCKLE and WOTHKE (1999); BACKHAUS et al. (2006)). In this context, it is important to note that in SEM all Likert-type scaled variables are treated as if they were continuous. This prevailing practice is justified by the assumption that each of the observed categorical variables is the expression of an underlying (and hence unobserved) counterpart that is of continuous scale. In addition, the underlying continuous variables are assumed to be multivariate normally distributed (BENTLER and WU (1995)). For the estimation of the initial model as depicted in Fig. 5-3 the ML estimator will be used given its desirable properties, such as inferential statistics for model modification. The sample size of n = 200 for the initial model constitutes no restriction for the application of the ML estimator as can be seen from Table 5-2.

However, given the distributional properties of the observed variables as discussed in chapter 3, a closer look has to be taken into the issue of multivariate normality of the input data. Quite a few of the variables presented in Table 5-1 show distributional properties that correspond with the findings of POATE and DAPLYN (1993) who stated that *"...data from small-holders are typically skewed with a concentration of low value observations and a few high value outliers in the long right tail"*. This finding applied to the SEM context in essence suggests a nonlinear transformation of the respective data. Following WEST et al. (1995) who outlined remedies for structural equation modeling with non-normally distributed data, the variables of the initial model that were described in chapter 3 as being skewed, were treated by logarithmic transformation. This was done based on a prior Kolmogorov-Smirnov Test for normal distribution.[62]

62 The Kolmogorov-Smirnov test procedure was used to test the null hypothesis that a sample comes from a particular distribution, in this case a normal distribution (see SIEGEL and CASTELLAN (1988)).

The estimation procedure of the initial model as depicted in Fig. 5-3, using the Maximum-Likelihood discrepancy function of AMOS®, results in a local minimum, which confirms that the estimation process yields an admissible solution. The next steps are then to (i) assess the individual model parameters and subsequently (ii) to assess the model as a whole. Following the systematic model assessment strategy as outlined by BYRNE (2001), firstly the parameters are evaluated under three different aspects: (i) the feasibility of the parameter estimates, (ii) the appropriateness of the standard errors, and (iii) the statistical significance of the parameter estimates.

With regard to the feasibility of parameter estimates, the sign and size of the parameters are of particular interest. Ideally, parameter sign and size are consistent with the model underlying theory. Table 5-3 presents the regression weights, i.e., the factor loadings of the initial model.

Table 5-3 Initial Model - Regression Weights

			Estimate	S.E.	C.R.	P
educ	<---	pref	1			
age	<---	pref	-0,346	0,072	-4,82	***
gndr	<---	pref	0,286	0,09	3,194	**
s_exp	<---	rend	0,433	0,103	4,2	***
a_hh	<---	rend	0,548	0,131	4,193	***
dsct	<---	rend	-0,007	0,03	-0,252	0,801
a_li	<---	rend	1			
madi	<---	minc	1			
incm	<---	minc	-4,446	1,605	-2,77	**
cred	<---	minc	-1,759	0,547	-3,213	**
prce	<---	minc	1,381	0,433	3,191	**
soil	<---	biof	1			
slope	<---	biof	1,02	0,057	17,994	***
fasi	<---	biof	0,85	0,047	18,071	***
rest	<---	unce	24561,623	22476958,7	0,001	0,999
resi	<---	unce	1			
flbr	<---	unce	-68,996	63140,641	-0,001	0,999
memb	<---	pisc	1,569	0,135	11,664	***
n_ext	<---	pisc	0,621	0,113	5,503	***
infc	<---	pisc	0,28	0,06	4,685	***
vsts	<---	pisc	1			
tist	<---	inpa	1			
inpa	<---	pref	0,132	0,13	1,016	0,309
inpa	<---	rend	0,027	0,035	0,789	0,43
inpa	<---	minc	-0,01	0,08	-0,121	0,904
inpa	<---	biof	-0,03	0,052	-0,579	0,563
inpa	<---	unce	-2027,956	1855833,74	-0,001	0,999
inpa	<---	pisc	0,138	0,06	2,315	*

* indicates significance at the .05 level; ** at the .01 level; *** at the .001 level

As discussed earlier, some of the estimates (in the 2nd column) have been fixed to 1 to determine the scale of the latent constructs as outlined in chapter 5.2. Without going into detail, one can see at a glance that some of the regression paths seem to be unreasonable, e.g., the path going from the latent variable for risk and uncertainty (*unce*) to the observed variable for tenure security (*rest*). Additionally, the paths from *unce* to the observed variable of labor availability in the extended family (*flbr*) and respectively the path from *unce* to the latent construct for the inclination to participate (*inpa* ← *unce*) seem to be difficult to deal with.

Furthermore, the standard errors in the 3rd column are extremely large for the respective paths indicating that the parameters can not be determined with sufficient accuracy (JÖRESKOG and SÖRBOM (1989)). If such values are encountered it is advisable to additionally have a closer look on the variances of the error terms, the variances of the latent variables as well as on the correlations of the exogenous factors before further interpreting results. If estimation results for correlations greater than 1 and/or negative variances are encountered this is a clear sign to discard the initial model. However, the estimated factor correlations as outlined in Table 5-4 do not result in values greater than 1, which is an indication that the model needs some improvement but is not to be completely discarded.

Table 5-4 Initial Model – Exogenous Factor Correlations

			Estimate
pref	<-->	rend	0,082
rend	<-->	minc	-0,375
minc	<-->	biof	-0,049
biof	<-->	unce	-0,267
pisc	<-->	unce	0,057
pref	<-->	minc	-0,165
rend	<-->	biof	0,402

minc	<-->	unce	-0,1
pisc	<-->	biof	0,179
pref	<-->	biof	0,029
rend	<-->	unce	0,196
minc	<-->	pisc	-0,276
pref	<-->	unce	0,065
rend	<-->	pisc	0,362
pref	<-->	pisc	0,017

Likewise, the variances of the error terms (*e1* – *e22*), the variance of the residual error (*res1*), and the variances of the latent exogenous variables do not show negative estimates.[63] Thus, the correlations as well as the variances of the initial model as shown in Table 5-4 and Table 5-5 do not directly contradict the feasibility of the parameter estimates.

63 Four error variances were fixed to zero assuming that the respective manifest variable can be measured with sufficient accuracy, e.g., the binary variable TIST membership or variables with a high degree of information and precision of the measured variable.

Table 5-5 Initial Model - Variances

	Estimate	S.E.	C.R.	P
pref	0,069	0,007	9,975	***
rend	2,517	1,152	2,185	0,029
minc	0,25	0,157	1,598	0,11
biof	0,729	0,088	8,287	***
unce	0	0	0,001	1
pisc	0,404	0,08	5,035	***
res1	0,225	0,023	9,946	***
e1	0			
e2	0,084	0,008	10,015	***
e3	0,111	0,011	9,975	***
e4	0,265	0,056	4,692	***
e5	0,595	0,1	5,955	***
e6	0,999	0,1	9,975	***
e7	19,488	2,023	9,631	***
e8	4,457	0,448	9,938	***
e9	32,575	3,302	9,865	***
e10	0,221	0,057	3,877	***
e11	0,04	0,033	1,223	0,221
e12	0,159	0,028	5,673	***
e13	0,231	0,033	7,078	***
e14	0,157	0,022	7,011	***
e15	0			
e16	0,819	0,082	9,975	***
e17	0,205	0,02	10,117	***
e18	0			
e19	0,826	0,083	9,997	***
e20	0,258	0,026	10,02	***
e21	0,591	0,059	9,975	***
e22	0			

Taking into account the aforementioned excursus on the correlations and the variances of the initial model, next, the statistical significance of the parameter estimates is going to be discussed. The test statistic of interest with regard to parameter significance is the critical ratio (C.R.) as shown in the 4th column of Table 5-3. The critical ratio is the quotient of the parameter estimate divided by its standard error. In essence, the C.R. works like a z-statistic that tests the parameter estimate to be different from zero. The null hypothesis H_o in the case of parameter estimation sets the parameter equal to zero. If C.R. >± 1.96, H_o can be rejected based on a .05 significance level, marked as * in Table 5-3. It can be seen that the majority of the observed variables loads statistically significant (at the .01 level or even at the .001 level) on their underlying constructs. In addition, the respective signs are mostly expected and consistent with the theory as outlined in chapter 4.1 and 4.2.

However, some unexpected results become evident as well, e.g. it is not obvious that greater labor availability in the household as indicated by the number of descendants working in agricultural production (*dsct*) loads negatively on the resource endowments of the household. Theoretically, one would have expected a positive sign at this point. Given some insignificant parameter estimates, the above mentioned regression paths that seem to exhibit estimates out of proportion and standard errors that are excessively large for some paths in the initial model, it is of primary interest to locate the source of misfit and to re-specify the model subsequently. The source of model misfit can be located by considering the overall model fit more closely, which is the topic of the subsequent chapter 5.4. Then, chapter 5.5 discusses suggestions for model modification. The aim is to come up with a model that is well fitting on the one hand and has sufficient explanatory potential on the other.

5.4. Model Testing and Test Statistics

Going back to equation (5.1), the null hypothesis (H_0) being tested in SEM states the following. The covariance matrix of the observed variables is a function of a set of parameters. If the established model were correct and its parameters known, the population covariance matrix would be exactly reproduced. As discussed, $\Sigma(\theta)$ stands for the restricted covariance matrix as specified by the hypothesized model and S is the sample variance-covariance matrix. The aim of the estimation process is to apply one of the discrepancy functions outlined in 5.3 and to fit a model with a residual matrix as small as possible, i.e., [$S - \Sigma(\theta)$ = minimal]. Discussing the issue of inferential statistics (see Table 5-2) the χ^2 value was introduced as a measure for the difference between the sample variance-covariance matrix and the reproduced variance-covariance matrix of the model. However, chi-square computed with a probability level and usually displayed with degrees of freedom serves only as a rapid model appraisal tool. For the initial model, the chi-square was 297,360 with 186 degrees of freedom and a probability level of .000.

As discussed in chapter 5.1, the aim in SEM is *not* to reject the null hypothesis as it is intended that the postulated model is true for the population. Thus, based on only this analysis a preliminary conclusion is that the data does not fit the hypothesized model entirely adequate. However, the mere testing for statistical significance is rather atypical when referring to covariance structures as it is mainly driven by degrees of freedom which result from the number of data points available (i.e., elements in the sample covariance matrix) and the number of parameters to be estimated (BENTLER and BONETT (1980)). In addition, there is an ongoing debate about the usefulness and feasibility of statistical significance testing in SEM.[64] This fact led to the development of an array of more 'applied' model fit indices which are also reported in AMOS®. These fit indices are usually used as adjuncts to the χ^2 statistic. Table 5-6 presents the most important model fit indices in structural equation modeling, which will then be discussed in the following, based on the results of the initial model.

Table 5-6 SEM - Model Fit Criteria

Criterion	Value Range	Interpretation	Reference
Chi-square	Tabled χ^2 value	Comparison of obtained χ^2 value with tabled value for given df	
Normed chi-square (CMIN/DF)	1.0 – 5.0	Less than 1.0 poor model fit; more than 5.0 indicates a need for improvement	WHEATON et al. (1977)
Root-mean-square-residual (RMR)	Researcher defines level	Indicates the closeness of Σ to S matrix	JÖRESKOG and SÖRBOM (1984) HU and BENTLER (1995)
Goodness-of-fit (GFI)	0 (no fit) to 1 (perfect fit)	Value close to .95 reflects a good fit	HU and BENTLER (1995)
Adjusted GFI (AGFI)	0 (no fit) to 1 (perfect fit)	Value adjusted for df, with .95 for a good model fit	HU and BENTLER (1995) FAN et al. (1999)
Parsimony-goodness-of-fit index (PGFI)	lower values than threshold level generally perceived as 'acceptable' for other normed fit indices	addresses the issue of parsimony in SEM - nonsignificant $\chi2$ statistics and GFI & AGFI statistics in the .90s, accompanied by PGFI in the .05s, not unexpected	JAMES et al. (1982) MULAIK et al. (1989)
Normed fit index (NFI)	0 (no fit) to 1 (perfect fit)	Value close to .95 indicates a good model fit – tends to underestimate fit in small samples	BENTLER and BONETT (1980) LOEHLIN (2004)
Comparative fit index (CFI)	0 (no fit) to 1 (perfect fit)	Value close to .95 indicates a good model fit. Revised form of NFI, taking the sample size into account.	BENTLER (1990)

64 BYRNE (2001) even reports the Board of Scientific Affairs for the American Psychological Association to work on a broader review to study the feasibility of phasing out the use of traditional null hypothesis testing.

Tucker-Lewis index (TLI)	0 (no fit) to 1 (perfect fit)	Value close to .95 indicates a good model fit	TUCKER and LEWIS (1973)
Parsimonious fit index (PRATIO, PNFI, PCFI)	0 (no fit) to 1 (perfect fit)	Comparison of values in alternative models	JAMES et al. (1982) MULAIK et al. (1989)
Noncentrality parameter (NCP)	The smaller λ, the better the model fit	Noncentrality parameter λ depicts the discrepancy between Σ and Σ(θ)	BOLLEN (1989) HU and BENTLER (1995) SATORRA and SARIS (1985)
Root-mean-square error of approximation (RMSEA)	<.05 (.08 <RMSEA<.10 mediocre fit and RMSEA>.10 poor fit)	Values less than .05 indicate a good model fit	BROWNE and CUDECK (1993)
Akaike information criterion (AIC)	0 (perfect fit) to positive values	Compares values in alternative models	AKAIKE (1987)
Expected cross-validation index (ECVI)	no predetermined range of values	ECVI compares models – the model with the smallest ECVI has the greatest potential for replication	BROWNE and CUDECK (1989)
Hoelter	value >200 desirable	Adequacy of sample size is assessed	HOELTER (1983)

Source: SCHUMACKER and LOMAX (2004) – modified; BYRNE (2001) and additional references as indicated in the table

By default, all of the above outlined fit statistics are reported for three comparative models: the actually specified (or default) model, the saturated, and the independence model. The saturated and the independence model are two extremes with the actual model lying somewhere in between the two. In the saturated model, the number of data points available for estimation equals the parameters to be estimated. Chi-square is zero in the saturated model and the model is least restricted. In the independence model on the other hand, there is complete independence of all variables (i.e., no correlations amongst variables), chi-square takes a maximum value, and the model is thus most restricted (see Table 5-7).

Table 5-7 outlines the number of parameters to be estimated in the initial model (NPAR) and CMIN which stands for the discrepancy between *S*, the unrestricted sample covariance matrix and Σ(θ), the model implied covariance matrix.

Table 5-7 Initial Model: Model Fit Summary: CMIN

Model	NPAR	CMIN	DF	P	CMIN/DF
Default model	67	297,360	186	,000	1,599
Saturated model	253	,000	0		
Independence model	22	1724,945	231	,000	7,467

Thus, in effect this value is equal to the likelihood ratio test statistic χ^2. As BOLLEN (1989) points out, for large samples the CMIN statistic has a central χ^2 distribution with the following degrees of freedom:

$$\tfrac{1}{2}\,(p)(p+1)-t \qquad (5.29),$$

where p is the number of observed variables, and t the number of parameters to be estimated. The CMIN value tests H_0 [$\Sigma = \Sigma(\theta)$]. As BOLLEN (1989) shows the null hypothesis can be written as $\Sigma - \Sigma(\theta) = 0$ and thus χ^2 can be seen as statistic that tests the extent to which all residuals in $\Sigma - \Sigma(\theta)$ are zero. The DF column in Table 5-7 reports the degrees of freedom of the model and the P (probability) value related to CMIN indicates the likelihood to obtain a CMIN value that exceeds the CMIN value in case H_0 is true. Therefore, the higher the P value of CMIN, the closer the fit between the hypothesized model (under H_0) and the perfect fit (BOLLEN (1989)). As mentioned above, solely taking into account this information our present model should be rejected. However, in interpreting this test statistic one has to bear in mind that it is (i) sensitive to sample size (see MACCALLUM et al. (1996) for a more detailed discussion) and (ii) based on the assumption of a central χ^2 distribution, assuming that H_0 is correct. Hypothesized models that fit the data well only based on the χ^2 statistic have been very rare in most empirical SEM research (JÖRESKOG and SÖRBOM (1996); MACCALLUM et al. (1996); BYRNE (2001)). Taking this limitation of the χ^2 (CMIN) statistic into account, the other goodness-of-fit indices for the initial model are examined more closely in the following.

One of the early-developed statistics to address model fit given the discussed shortcomings of the chi-square statistic was the normed chi-square (reported as CMIN/DF above) of WHEATON et al. (1977). The value of the initial model (1,599) lies between the two-sided thresholds established by WHEATON et al. (1977), stating that a value less than 1.0 indicates a poor model fit while a value greater than 5.0 also indicates the need for model improvement.

Table 5-8 shows four more applied indices of model fit: the root mean square residual (RMR), the goodness-of-fit index (GFI) together with its degrees of freedom adjusted counterpart, the AGFI, and the parsimony goodness-of-fit index (PGFI). The root mean square residual (RMR) is defined as the average residual value resulting out of fitting the model implied covariance matrix $\Sigma(\theta)$ to S, the covariance matrix of the sample data (JÖRESKOG and SÖRBOM (1984)).

Table 5-8 Initial Model: Alternative Fit Indices: RMR, GFI, AGFI, PGFI

Model	RMR	GFI	AGFI	PGFI
Default model	,280	,876	,832	,644
Saturated model	,000	1,000		
Independence model	,375	,536	,491	,489

As Table 5-6 points out, the researcher has to define a critical value range by oneself as the residuals are relative to the sizes of the variances and covariances of the observed variables. If the RMR value is to be interpreted this is usually done in its standardized form, i.e. in the metric of the *correlation* matrix (HU and BENTLER (1995)). This standardized RMR is the average value across all standardized residuals with a range from zero to one. The smaller the RMR, the better. An RMR of zero would indicate a perfect model fit (ARBUCKLE and WOTHKE (1999)). Comparing the RMR of the initial model with results presented in WHEATON et al. (1977) the figure of RMR = .280 is in the range of expected values. WHEATON et al. (1977) accept a model with an RMR of .284.

However, simultaneously other fit indices should be considered, namely, the goodness-of-fit index (GFI) as a measure for the relative amount of variance and covariance in S that is jointly explained by Σ (JÖRESKOG and SÖRBOM (1984)). The AGFI is almost the same measure but modifies GFI with respect to the degrees of freedom. It therefore addresses the issue of model parsimony by penalizing for the inclusion of additional parameters. Table 5-6 shows both indices to range from zero to one. The closer the value to one, the better the actual model fit. The values of the initial model (.876 for the GFI and .832 for the AGFI) do not completely satisfy the threshold set by HU and BENTLER (1995) and FAN et al. (1999) who stated that in both cases a value close to .95 would reflect a good model fit. Still, the GFI of .876 indicates that 88% of the variance and covariance in S is predicted by the model implied variance-covariance matrix. The last fit index presented in Table 5-8 is the parsimony goodness-of-fit index (PGFI). The PGFI addresses the complexity of the hypothesized model (i.e., number of parameters to be estimated) and thus combines information on model fit and model parsimony in one single index. The PGFI in general has a lower threshold level than the GFI and the AGFI. MULAIK et al. (1989) considered a situation with a non-significant χ^2 statistic, GFI and AGFI statistics in the area of .90 and a PGFI value close to .05 as "*not unexpected*". The PGFI of the initial model (.644) is therefore coherent with the previously presented GFI and AGFI values.

The next group of fit indices shown in Table 5-9 compares the proposed model with the independence model in order to evaluate the fit of the former. The indices presented are the normed fit index (NFI) of BENTLER and BONETT (1980), the Tucker-Lewis index (TLI), first presented by TUCKER and LEWIS (1973) and the comparative fit index (CFI) (BENTLER (1990)).

Table 5-9 Initial Model: Model Comparison: NFI, TLI, CFI

Model	NFI	TLI	CFI
Default model	,828	,907	,925
Saturated model	1,000		1,000
Independence model	,000	,000	,000

The NFI index basically rescales the chi-square value into a measure ranging from zero (no model fit at all) to one (perfect fit of the model) (BENTLER and BONETT (1980)). However, especially in smaller samples the NFI was reported to underestimate model fit (BYRNE (2001)). To address this issue BENTLER (1990) proposed the comparative fit index (CFI), a revised form of the NFI which takes sample size into account. Originally, values for both, the NFI as well as the CFI greater than .90 were considered to be indicative for a good model fit (BENTLER (1992)). But more recent literature suggests to set the threshold to .95 (HU and BENTLER (1999)). The NFI of the initial model is lower than threshold requirements while the CFI of .925 is higher than the originally proposed threshold of .90 and draws near the revised cutoff point of .95. The Tucker-Lewis index (TLI) was originally developed for factor analysis as a reliability coefficient for maximum likelihood estimation procedures and then enhanced to assess model fit in structural equation modeling (TUCKER and LEWIS (1973)). The value range of the index is consistent with the other fit indices presented here. Values close to zero point toward no model fit while a value of one indicates a perfect model fit. The value of .907 for the initial model presented in Table 5-9 is a good starting point for further model improvement. It meets the threshold of .90 but fails to meet the stricter threshold of .95, which applies especially to larger samples.

Table 5-10 below presents parsimony-adjusted measures for the normed fit and the comparative fit index, which are respectively the PNFI and the PCFI value.

Table 5-10 Initial Model: Parsimony-Adjusted Measures: PRATIO, PNFI, PCFI

Model	PRATIO	PNFI	PCFI
Default model	,805	,666	,745
Saturated model	,000	,000	,000
Independence model	1,000	,000	,000

In addition, PRATIO is the acronym for the parsimony-ratio index which was introduced by JAMES et al. (1982). In principle, these indices address another time the issue of model parsimony (as described above for the AGFI and the PGFI), i.e. the number of estimated parameters needed to achieve a desired level of model fit is taken into account. Again, the value range for the indices is standardized from zero to one whereas a value close to one indicates superior model fit than lower values close to zero. As the NFI value in Table 5-9 was already lower than the CFI value, it is not unexpected that the PNFI and the PCFI show the same relationship. The actual values of the three indices are not surprising and will have to be compared to their respective equivalents after the model has undergone modifications. This will be discussed in the next chapter. With regard to the issue of model parsimony, another group of indices presented in Table 5-12 is of interest. The Akaike information criterion (AIC) assesses model fit and model parsimony (AKAIKE (1987)). In AMOS® the value is calculated as $\chi^2 + 2q$ (5.30), with q being the number of parameters in the model. Usually, the value of equation (5.30) is positive and the closer the value to zero the more parsimonious the model (ARBUCKLE and WOTHKE (1999)). The CAIC index in the fifth column of Table 5-11 is the so called consistent version of Akaike's information criterion which also takes sample size into account (BOZDOGAN (1987); BANDALOS (1993)). Both indices compare different models with each other whereas smaller index values indicate a better model fit.

Table 5-11 Initial Model: Akaike Information Criterion: AIC

Model	AIC	BCC	BIC	CAIC
Default model	431,360	448,871	652,347	719,347
Saturated model	506,000	572,125	1340,474	1593,474
Independence model	1768,945	1774,695	1841,508	1863,508

The BCC and BIC indices in the third and fourth column of Table 5-11 are both different versions of the Akaike information criterion but 'penalize' more severely for model complexity (BROWNE and CUDECK (1989)). Amongst all parsimony-based fit indices WILLIAMS and HOLAHAN (1994) found the AIC to perform best. The main finding of the values presented in Table 5-11 is that for all four fit indices the default (i.e., hypothesized) model values are *smaller* than the values of the saturated and the independence model which is actually a desired and intended finding.

The next model fit index discussed is the noncentrality parameter (NCP). As outlined above, BOLLEN (1989) shows that the CMIN statistic has a central χ^2 distribution for large samples. In case the hypothesized model is incorrect, the χ^2 distribution is no longer central but has to be approximated by the noncentrality parameter, λ (HU and BENTLER (1995)). The noncentrality parameter λ measures the discrepancy between Σ and $\Sigma(\theta)$. In doing so, greater values of λ are an evidence for a greater discrepancy between Σ and $\Sigma(\theta)$, or, the smaller the value of the NCP the better the model fit. In essence, the central χ^2 distribution is a special case of the noncentral χ^2 distribution for λ being equal to zero (STEIGER (1990)). In addition, STEIGER (1990) established confidence intervals for the NCP in order to address its precision.

Table 5-12 Initial Model: Noncentrality Parameter: NCP

Model	NCP	LO 90	HI 90
Default model	111,360	68,250	162,392
Saturated model	,000	,000	,000
Independence model	1493,945	1365,643	1629,684

Table 5-12 shows a NCP of 111,360 for the initial model, which is equal to the χ^2 value of 297,360 minus the degrees of freedom (DF = 186). The confidence interval assures that the true population value for NCP lies in 90% between the lower 68,250 and the higher 162,392 bound.

The root mean square error of approximation (RMSEA) presented in Table 5-13 addresses the following question: *"How well would the model, with unknown but optimally chosen parameter values, fit the population covariance matrix if it were available?"* (BROWNE and CUDECK (1993)). BYRNE (2001) reports this index to be recognized in the literature as one of the most informative indicators to assess model fit. It is responsive to model complexity (i.e., number of estimated parameters) as it is expressed by degrees of freedom. Values below .05 represent a good model fit, values between .08 and .10 are reported as 'mediocre fit' while values greater than .10 indicate a poor model fit (BROWNE and CUDECK (1993); MACCALLUM et al. (1996)). In the third and fourth column of Table 5-13 a 90% confidence interval is reported around the RMSEA point estimate of .055 for the initial model.

Table 5-13 Initial Model: Root Mean Square Error of Approximation: RMSEA

Model	RMSEA	LO 90	HI 90	PCLOSE
Default model	,055	,043	,066	,240
Independence model	,180	,172	,188	,000

In interpreting the overall model fit not only the point estimate for RMSEA has to be taken into account but also its confidence interval. The wider the confidence interval the more imprecise the point estimate for RMSEA. PCLOSE in the fifth column of Table 5-13 tests for the closeness of fit, i.e. test of the hypothesis that RMSEA is below .05 in the population (BROWNE and CUDECK (1993)). Preferably the value of PCLOSE should be greater than .50 (BYRNE (2001)). The RMSEA value for the initial model is .055 with the 90% confidence interval from .043 to .066 and the closeness of fit statistic (PCLOSE) being .240. The narrow confidence interval indicates a good precision of the point estimate of RMSEA (= .055). The upper bound of the confidence interval is still close to .06 which was suggested by HU and BENTLER (1999) to be indicative of a good model fit. However, the probability level of the test of close fit is lower than the proposed threshold, which again suggests that the initial model must not be completely rejected but potential misspecifications should be detected in a next step in order to improve the model.

The last two fit statistics to be discussed are the expected cross-validation index (ECVI) and Hoelter's critical *N* (CN). Table 5-14 presents the ECVI for the initial model (value = 2.168) which is a measure to assess the likelihood that the hypothesized model (for a single sample) cross-validates in similar sized samples to be taken from the same population (BROWNE and CUDECK (1989)).

Table 5-14 Initial Model: Expected Cross Validation Index: ECVI

Model	ECVI	LO 90	HI 90
Default model	2,168	1,951	2,424
Saturated model	2,543	2,543	2,543
Independence model	8,889	8,244	9,571

The ECVI measures the discrepancy between the fitted covariance matrix in the analyzed sample and the expected covariance matrix in a sample of similar size from the same population. As Table 5-6 points out there is no predetermined range of values for the ECVI as it is a tool for model comparison. In comparing different models, the model with the smallest ECVI has the greatest chance for replication. In comparison to the saturated and the independence model the initial model ('default model' in Table 5-14) has the lowest ECVI, meaning that it represents the best fit to the data within the continuum of the three models. Again, a confidence interval is provided to assess the precision of the estimated ECVI value. For the initial model, the confidence interval goes from 1,951 to 2,424 and is narrower than the confidence interval of the independence model thus indicating that the point estimate for the ECVI of the initial model is already a reasonable approximation to the population.

Finally, Hoelter's critical *N* addresses the issue of model fit slightly different. It assesses the adequacy of the sample size in order to get a satisfactory model fit for a χ^2 test (HOELTER (1983); HU and BENTLER (1995)).

Table 5-15 Initial Model: Hoelter's Critical N

Model	HOELTER .05	HOELTER .01
Default model	147	157
Independence model	31	33

A value for both, the Hoelter .05 and the Hoelter .01 statistic that is greater than 200 was originally recommended in order to assume that the model adequately represents the sample data (HOELTER (1983)). According to HOELTER (1983), the interpretation of the .05 statistic of Table 5- is as follows. The initial model would have been accepted at the .05 level for the given sample moments if the sample size had been n = 147. As the sample size of the initial model (N = 200) exceeds this value the model has to be rejected from Hoelter's point of view. For the initial model, both Hoelter statistics do not meet the desired threshold of 200 suggesting that the sample size of the initial model might be rather at the lower margin. However, other authors have put this threshold into question arguing that it might be chosen overly conservative (see ARBUCKLE and WOTHKE (1999)).

5.5. Model Modification

In summary, the preceding chapter 5.4 showed the initial model hypothesized in Fig. 5-1 to be a good starting point to explain the inclination of smallholders to participate in the TIST program. However, some of the fit indices discussed above suggest modifying the initial model in order to enhance its explanatory power. It is important to note that the strictly confirmatory approach of SEM is abandoned once the model is modified (i.e., re-specified). In Fig. 5-1 the modification process is depicted by the feedback loop between the two rectangles of 'model testing/evaluation' and 'model specification'.

As mentioned above, the goal of the modification process is twofold. On the one hand potential sources accounting for the misfit of the model should be located. On the other hand the model should be re-specified accordingly. Finally, the goal is to find a substantively meaningful and statistically well-fitting model. Note that a modification process is not only justified from a practical point of view given the costs associated with field research and questionnaire-based data collection but has its theoretical foundations in the general strategic framework for testing structural equation models as presented by (JÖRESKOG (1993)). As discussed on the outset of this chapter he introduced the differentiation between a strictly confirmatory strategy, the strategy of alternative models, and a model generating approach in SEM. The aim of chapter 5.5 is to modify the initial model according to different types of information available to detect model misspecification and to finally present an alternative, modified and better fitting model.

A first starting point is to reconsider the statistical significance of the estimated parameters in the model. SCHUMACKER and LOMAX (2004) suggest fixing parameters that are not statistically significant, i.e., have small critical ratios (C.R. values in Table 5-3) to zero in a subsequent model. However, they caution that statistical significance is related to sample size and therefore parameters that are not significant in small samples can become significant in a larger one. To stress in addition, theoretical deliberations must at all times guide the re-specification process. A non-significant parameter that is of substantive theoretical interest should remain in the model anyhow. In reviewing the parameter estimates and critical ratios of Table 5-3 the following regression paths in the measurement models are found to be non-significant: *rend* → *dsct* indicates that the labor availability in the household measured by the number of descendants (*dsct*) might not be an adequate measure to determine the overall resource endowment of the household. In addition, the sign of the regression path is rather unexpected. Why should greater labor availability load negatively on the household's resource endowment? There is no obvious theoretical foundation for such a finding.

Furthermore, the path going from the latent variable for risk and uncertainty (*unce*) to the observed variable for tenure security (*rest*) is firstly not significant and second, the standard error (3rd column) for the respective path is extremely large which indicates that the parameter can not be determined with sufficient accuracy. The third path in the measurement models of Fig. 5-3 found to be non-significant is from labor availability in the extended family (*flbr*) to the construct describing risks and uncertainties related to the adoption of a new technology (*unce*). Again, the parameter estimate is insignificant, has a small critical ratio (C.R.) but a large standard error, which makes the estimate unreliable. Thus, the specification of all three measurement variables (*dsct*, *rest*, *flbr*) is to be reconsidered in a subsequent model. The variable measuring access restrictions of the respective household (*rest*) is of most interest amongst the three aforementioned variables due to the theoretical deliberations made in earlier chapters. Therefore, an effort shall be made to retain this variable in the model as long as possible (i.e., not to fix the parameter to zero).

Another source of information with regard to potential model modifications is the squared multiple correlation. In factor analysis, squared multiple correlation is a measure of reliability for the observed variables. A variable's squared multiple correlation is the proportion of its variance that is accounted for by its predictors (ARBUCKLE and WOTHKE (1999)). The squared multiple correlation is usually independent of all units of measurement. Thus, for the observed variables squared multiple correlations are scaled from zero to one and indicate how well the observed variables serve as measures for the latent variables. Squared multiple correlations also exist for the structural equations of the model. In this case, they specify the strength of the structural relationships between the latent variables (scaling from zero to one). The calculated squared multiple correlations for all measured variables as well as for the latent endogenous variable *inpa*, the household's inclination to participate in the TIST project are reported in Table 5-16.

As can be seen in Table 5-16 some observed variables are statistically rather weak measures for their underlying hypothesized constructs. The estimate for *flbr* and *dsct* for example is equal to zero suggesting that those measures might either proxy for another latent construct or have to be deleted in a modified model version. BACKHAUS et al. (2006) suggest a cutoff value of 0.4 for the squared multiple correlations of the indicator variables. Smaller values indicate a need for model re-specification. For the initial model, the squared multiple correlations of ten observed variables of the model's measurement part do not meet this criterion.

Table 5-16 Initial Model: Squared Multiple Correlations

	Estimate
inpa	0,099
tist	1
rest	1
resi	0
flbr	0
soil	0,821
slope	0,767
fasi	0,77
vsts	0,406
memb	1
n_ext	0,159
infc	0,11
prce	0,923
madi	0,053
incm	0,132
cred	0,778
a_li	0,114
s_exp	0,641
a_hh	0,559
dsct	0
educ	1
age	0,09
gndr	0,049

The next valuable information in the attempt to modify the initial model is the residual matrix. The residual matrix can point to general model misspecification. It describes the discrepancy between the model-implied covariance matrix $\Sigma(\theta)$ and the sample covariance matrix S. The elements of the residual matrix are the differences between the covariances in $\Sigma(\theta)$ and S which means that for each pair of observed variables there is one residual in the residual matrix (JÖRESKOG (1993)). The elements of the residual matrix are dependent on the measurement unit of the respective observed variable, which makes them difficult to interpret. However, the values of the residual matrix should in general be small in magnitude and all elements should approximately lie within the same value range, i.e., values should not be larger for one variable than for another. In general, large values in the residual matrix point to misspecification of the entire model while large values for single variables indicate that the respective variable might be mis-specified (SCHUMACKER and LOMAX (2004)). In App. 1 the residual matrix for the initial model is shown. At a first glance, no specific problem with regard to the overall magnitude of the matrix elements can be detected. Two individual value outliers might exist, namely (i) the residual covariances between *madi* (i.e., the distance to the next marketplace to buy and sell agricultural goods) and the livestock asset variable of the household (*a_li*), and (ii) the latter variable and the variable for off-farm income (*incm*). The respective variables will have to be reconsidered when attempting to modify the initial model.

As mentioned earlier, the residual matrix elements depend on the measurement unit of their observed variables, which causes difficulties in interpretation. Therefore, the standardized values of the residual covariances are typically more suitable for model modification purposes. In the standardized residual covariance matrix the elements of the residual matrix are divided by an estimate of their standard error (JÖRESKOG and SÖRBOM (1984)). Thus in principle, the standardized residuals are interpreted as *z*-scores. As such, they represent estimates of the number of standard deviations the observed residuals differ from the zero residuals that would exist if the model would fit the data perfectly well. JÖRESKOG and SÖRBOM (1988) consider absolute values of the standardized residuals greater than 2.58 to be beyond a critical threshold. The residual matrix of the initial model containing 253 elements is presented in App. 2. In analyzing the standardized residual covariances it becomes clear that 16 out of the 253 elements (6%) are greater than the absolute value of 2.58, indicating that these particular covariances are not sufficiently well explained by the model.

Especially covariances of the labor availability in the extended family (*flbr*) with other observed variables like the farm size (*fasi*) or the age of the farmer (*age*) seem to be inadequately explained. There is an apparent accumulation of five (out of 16) values greater than 2.58 in the column of *flbr* which demonstrates a clear need to reconsider the specification (or removal) of this variable. Furthermore, it can be seen that two critical values exist for the standardized residual covariance of *resi* (i.e., the years of residence of the particular household in the community), namely with *flbr* and *dsct*. And for *infc*, which assesses all the sources, used by the farm household to get agricultural market and price information with two other observed variables, namely *s_exp* and *a_hh*. One critical standardized residual covariance exists for the following variables. Soil properties (*soil*) of the farm with *infc*; the size of the farm (*fasi*) with *dsct*; the variable *vsts* which delivers information on the connectedness of the farm household with household assets (*a_hh*); access to extension and training (*n_ext*) with *a_hh*; the price for capital (*prce*) with the livestock assets variable (*a_li*); and off-farm income (*incm*) with the sum of household expenditures (*s_exp*). A particularly high value exists for *dsct* and the age of the household head (*age*). Especially the aforementioned variables exhibiting large standardized residuals are looked at when the initial model is modified further down.

The fourth type of information related to potential model misspecification is the so called 'modification indices' (MIs) that can be requested in AMOS® together with the 'expected parameter change' (EPC) values. The modification indices capture potential misfit in the hypothesized model. JÖRESKOG and SÖRBOM (1988) introduce the MI statistic as a χ^2 statistic with one degree of freedom. It is important to note that this tool should only be used if the proposed model modification makes sense from a theoretical point of view. AMOS® provides MIs for fixed parameters in the model and indicates the expected decrease of the χ^2 statistic if this parameter would be freely estimated.[65] In general, MIs suggest possibilities to improve a model by increa-sing the number of parameters. It is obvious that this strategy will decrease the degrees of freedom of the model. The ultimate purpose is to modify the model according to selected MIs so that the value of the χ^2 statistic decreases faster than the degrees of freedom (ARBUCKLE and WOTHKE (1999)). The expected parameter change (EPC) values, labeled 'Par Change' in Table 5-17, indicate the expected parameter change (i.e., magnitude and direction) if the respective fixed parameter would be estimated freely in a re-specified version of the model (SARIS et al. (1987)). Thus, the MI values relate to the predicted improvement in the goodness of fit test while the EPC values yield information directly related to the model parameters.[66]

Table 5-17 Initial Model: Modification Indices Covariances

			M.I.	Par Change
e17	<-->	e16	9,322	0,047
e21	<-->	rend	10,313	-0,268
e21	<-->	e16	5,115	-0,104
e19	<-->	rend	5,153	0,213
e20	<-->	rend	8,376	0,149
e11	<-->	e19	5,718	0,054
e8	<-->	e18	4,833	0,224
e8	<-->	e20	4,866	-0,156
e9	<-->	rend	6,639	1,606
e9	<-->	e20	8,731	0,568
e10	<-->	e20	6,168	-0,044
e7	<-->	minc	9,51	0,463
e5	<-->	e21	7,148	-0,131
e5	<-->	e7	5,269	0,652
e6	<-->	e8	6,207	0,215
e2	<-->	e8	8,479	0,116
e3	<-->	res1	4,642	-0,024
e3	<-->	e22	4,642	-0,024
e3	<-->	e17	4,408	-0,013

Table 5-17 presents the MIs and EPC values for the initial model with regard to possible covariances. Only MI values greater than four are displayed in order to concentrate on parameters responsible for substantively meaningful potential model improvement. Of major interest in Table 5-17 are the covariances between error terms. All EPC values are fairly small. However, the error variances

65 All freely estimated parameters in the model have MIs equal to zero.

66 ARBUCKLE and WOTHKE (1999) note that in principle the change in the $\chi2$ statistic should approximate the MI value. However, the actual change of $\chi2$ can be even greater.

between e17 and e16 (*resi* and *flbr*), e9 and e20 (*incm* and *infc*), e5 and e21 (*a_hh* and *vsts*) and e2 and e8 (*age* and *madi*) are of possible interest as the predicted change in the goodness of fit test (χ^2) indicated by the MIs seems to carry some weight (MIs > 7). App. 3 shows the modification indices for the regression weights of the initial model. The most interesting paths are those indicating so called *cross-loadings*, i.e. paths going from a latent variable to a measurement variable that is hypothesized to already measure another factor. In App. 3 four cross-loading paths can be found: *rend* → *vsts*, *biof* → *infc*, *rend* → *infc*, and *inpa* → *gndr*. While the first three paths have rather high MIs arguing to specify additional paths between those variables, there is no theoretical background that would justify to free up those relationships. Neither is it theoretically sound to let the count of all the visits of different meetings in community groups (*vsts*) proxy for *rend*, the resource endowments of the farm, nor should the resource endowments of the farm be measured by *infc*, all the channels used to get agricultural market and price information. Furthermore, it seems not meaningful to let the latter proxy for the bio-physical characteristics of the farm (*biof*).

Of substantive interest however, especially from a theoretical point of view, is the suggested path from *inpa* to *gndr*. What this path does in essence is to point to the fact that the overall model fit would improve if a new path between the latent variable for the farmer's inclination to participate (*inpa*) and the variable measuring the gender aspect of the potentially participating household would be specified. This re-specification does not only make sense in the context of the model modification but also from a theoretical and empirical point of view. In theory, the gender of the head of the household often determines the disposition and preparedness of the household to adopt a new technology. Also empirically, gender matters very often in rural areas with regard to adoption decisions as female headed households struggling to integrate childcare, and farm work might simply lack capacities to give the new technology a try. Thus, in the modified version of the initial model the gender variable (*gndr*) is now directly measuring the inclination to participate (*inpa*) on the right hand side of the structural equation model in the measurement model for *inpa*. In addition, a new variable should be introduced in the measurement model for *inpa* which proxies for the religious affiliation or denomination of the household (head). This variable is labeled *reli* and measures whether or not the household belongs to the Anglican Church through which the TIST program was initially promoted.

In summarizing the different procedures used in the specification search, the following model modifications seem most appropriate. At first, the statistical significance of the estimated parameters in the model was reconsidered. Three regression paths performed poorly including the measurement variables *dsct*, *rest*, and *flbr*. Parameter estimates were non-significant, had small critical ratios, and in two cases (*rest* and *flbr*) a large standard error, which made the parameter estimate unreliable. As mentioned above, *rest*, the variable measuring access restrictions on the farm is considered of particular interest. Therefore, *rest* is kept in the modified model. However, *dsct* and *flbr* are deleted from the initial model given their weak regression path performance on the one hand, and on the other, their zero values with regard to the squared multiple correlation. As mentioned above, BACKHAUS et al. (2006) suggested a cutoff value of 0.4 for the squared multiple correlation of the indicator variables. Smaller values indicate a need for model re-specification. For the initial model, the squared multiple correlations of ten observed variables of the model's measurement part did not meet this criterion, amongst them *dcst* and *flbr* which are deleted in the subsequent model. The gender variable (*gndr*) also performed rather poor in the squared multiple correlation but is brought on the right hand side in the modified model, measuring now directly the inclination to participate in the program (*inpa*).

Furthermore, Table 5-16 shows low values for *madi* (i.e., the distance to the next marketplace to trade agricultural goods) and *infc* (information channels) which are also amongst the 'poor performers' in the residual matrix (*madi* in App. 1) and the standardized residual matrix (*infc* in App. 2). Both information taken together, the variables are dispensable in the modified model and therefore deleted. All the other variables add substantial explanatory power to the model and have been retained unchanged. At this point, no measurement error covariances were specified as suggested by the MIs of Table 5-17 because the respective variables are removed from the initial model. The graphical representation of the modified model is shown in the next chapter 5.6 together with the results of the re-specified model.

5.6. Model Results

The basic structure of the modified model presented in Fig. 5-4 is the same as for the initial model: the six latent variables *pref*, *rend*, *minc*, *biof*, *unce*, and *pisc* (left hand side of the model) are hypothesized to influence the household's inclination to participate, *inpa*, on the right hand.

The straight single-headed arrows going from the left to the right depict the hypothesized relationships. As in the initial model, these arrows are interpreted as six structural hypotheses depicting the assumed relationships between the latent variables on the left and the latent variable on the right. These inter-factor relations or regression structures amongst the latent variables are of central interest in the modified structural equation model. A separate measurement model specifies each latent variable on the left. The preferences of the household (*pref*) are now measured by education of the household head (*educ*) and the household head's age (*age*). The variable characterizing the household head's gender (*gndr*) was moved to the right hand side of the model due to the cross-loading from *inpa* on *gndr* shown in the modification indices for the regression weights of the initial model (see App. 3).

Now the gender variable is a direct measure for the inclination to participate (*inpa*). As in the initial model, *inpa* is also measured by actual program participation by the *tist* variable. In addition, a new variable is introduced in the measurement model for *inpa* which proxies for the religious affiliation or denomination of the household (head). The variable labeled *reli* measures whether or not the household belongs to the Anglican Church that originally promoted the TIST program. The resource endowments (*rend*) of the household are still measured by the income proxy sum of the household's general expenditures (*s_exp*) during the last year. Household assets (*a_hh*) monetarizing different kinds of assets necessary for daily life and agricultural production remain unchanged. The same is true for the variable measuring livestock ownership of the household (*a_li*).

Fig. 5-4 Modified Structural Equation Model

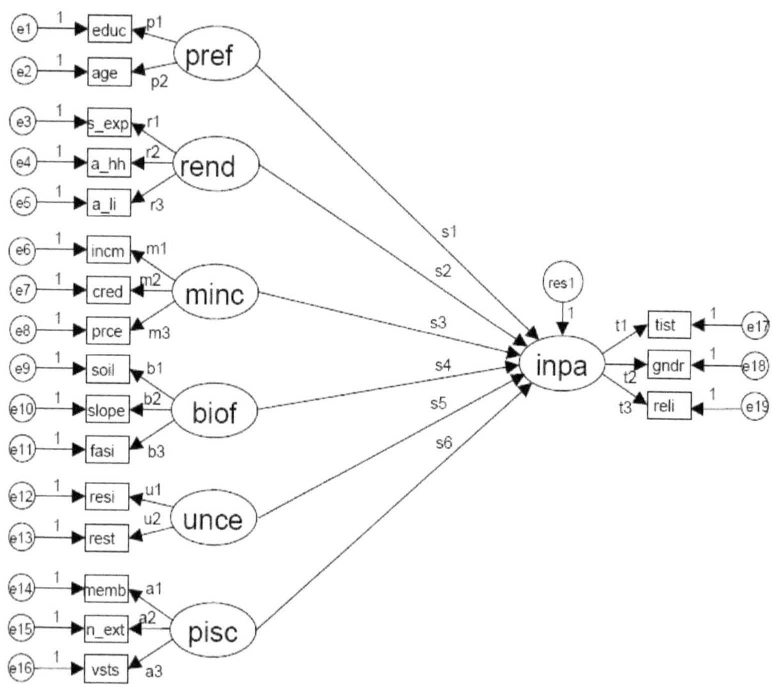

The fourth measured variable under resource endowments (rend) was the proxy for the labor availability within the household measured by descendants (dsct), able to assist in farming activities. As mentioned above, the poor regression weights and the zero values in the squared multiple correlation of dsct led to its deletion from the initial model. Market-based incentives (minc) in the modified model are measured by only three manifest variables, the off-farm income (incm), the availability of credit (cred), and the price proxy (prce). The variable measuring the distance to the next marketplace (madi) was removed due to the low squared multiple correlation value (see Table 5-16) and its poor performance in the residual matrix (see App. 1).

The latent variable for the biophysical characteristics of the farm (*biof*) remains completely unchanged with its three measurement variables for soil (*soil*), slope properties (*slope*), and the variable measuring the size of the farm (*fasi*). Risk and uncertainty the household faces (*unce*) is measured by only two manifest variables in the modified model as opposed to three in the initial model. The variable measuring the tenure security of the farmer (*rest*) was retained. In addition, *resi*, accounting for the years of residence of the particular household in the community remained in the modified model. The third manifest variable describing the overall labor availability in the extended family (*flbr*) was removed from the modified model due to poor squared multiple correlations and insufficiently explained covariances in the standardized residual covariance matrix (see App. 2). The construct for the perceived indicator of social capital (*pisc*) is now measured by household membership in different community organizations (*memb*), access to extension and training (*n_ext*), and actual meeting visits of different groups (*vsts*). The variable *infc* assessing all the sources for agricultural market and price information used by the farm household was removed from the initial model. This is mainly due to its cross-loadings on other latent constructs as shown in the modification indices for the regression weights of the initial model (see App. 3) and undesirably high values in the standardized residual covariance matrix (see App. 2).

Again, it is important to note that implicitly all the path coefficients going from the latent variables on the left to their respective indicators are all measurement hypotheses within the model. As above, all the indicator variables of the seven measurement models (six on the left and one on the right hand side) are influenced by measurement error as indicated by the associated error terms (*e1* – *e19*). Only *inpa* carries a residual error term (*res1*), which measures error inherent to the prediction of an endogenous latent variable by exogenous latent variables.

For the estimation of the modified model the Maximum-Likelihood discrepancy function was used. As the model reports to achieve a local minimum, the estimation process yields an admissible solution. Table 5-18 shows the regression weights (i.e., factor loadings) of the modified model that have to be compared with the initial model in the following.

First, the feasibility of parameter estimates, i.e., their sign and size are of particular interest. As in the initial model, some of the parameter estimates are fixed to 1 in order to determine the scale of the latent constructs (see chapter 5.2). While in the initial model some of the regression paths seemed to be unreasonable (e.g., path from *unce* to *rest*, path from *unce* to the observed variable of labor availability in the extended family (*flbr*) and path from *unce* to *inpa*) there are no such extreme parameter estimates in the modified model. Particularly three regression paths are now less difficult to interpret than in the initial model: *unce* → *rest*, *minc* → *incm* and *minc* → *prce*. The value of the parameter estimate and standard error of *unce* → *rest* significantly decreased. The value for the other two parameter estimates is more or less the same as in the initial model, while the respective signs changed which makes more sense from a theoretical point of view. A higher off-farm income reasonably measures greater market incentives and the higher the price for capital the lower the incentives of the market the farmer faces.

Table 5-18 Modified Model: Regression Weights

	Estimate	S.E.	C.R.	P
educ <--- pref	1			
age <--- pref	-0,339	0,074	-4,603	***
s_exp <--- rend	0,557	0,176	3,169	**
a_hh <--- rend	0,625	0,181	3,459	***
a_li <--- rend	1			
incm <--- minc	2,502	0,48	5,213	***
cred <--- minc	1			
prce <--- minc	-0,794	0,065	-12,227	***
soil <--- biof	0,961	0,055	17,354	***
slope <--- biof	1			
fasi <--- biof	0,853	0,049	17,424	***
resi <--- unce	1			
rest <--- unce	7,584	3,91	1,940	♦
memb <--- pisc	1,542	0,127	12,176	***
n_ext <--- pisc	0,633	0,112	5,646	***
vsts <--- pisc	1			
tist <--- inpa	1			
gndr <--- inpa	-0,104	0,047	-2,189	*
reli <--- inpa	0,232	0,063	3,711	***
inpa <--- pref	0,198	0,172	1,148	0,251
inpa <--- rend	0,098	0,124	0,788	0,431
inpa <--- minc	-0,007	0,056	-0,116	0,907
inpa <--- biof	-0,139	0,192	-0,726	0,468
inpa <--- unce	1,717	1,031	1,665	♦
inpa <--- pisc	0,136	0,064	2,11	*

* indicates significance at the .05 level; ** at the .01 level; *** at the .001 level, ♦ at the 0.1 level

Following the systematic model assessment strategy by BYRNE (2001) the appropriateness of the standard errors has to be assessed next. Looking at the 3rd column of Table 5-18, all standard errors are in a reasonable range of values. Solely, the standard error of the parameter estimate *unce* → *rest* is somewhat larger than the rest but far from the extreme values of the initial model. As mentioned above, if standard errors were extremely large this would indicate that the respective parameters can not be determined with sufficient accuracy (JÖRESKOG and SÖRBOM (1989)). However, this is not the case for the modified structural equation model.[67]

67 The additional examination of the variances of the error terms and the variances of the latent variables as well as on the correlations of the exogenous factors revealed no incon-

The third step is to look at the statistical significance of the parameter estimates. The critical ratio (C.R. in the 4th column of Table 5-18) tests the parameter estimates to be different from zero. Looking at the C.R. values for the modified model in Table 5-18 it is evident that the great majority of the observed variables in the measurement models load statistically significant (at the .01 level or even at the .001 level) on their underlying constructs. The path from *unce* to *rest* is significant at the 0.1 level in the respective measurement model. In the structural part of the modified model (i.e., relationships between the latent variables on the left and the latent variable on the right) the latent construct for the perceived indicator of social capital loads statistically significant (at the .05 level) on the inclination to participate in the TIST program. A similar effect has been observed in the initial model, which underlines the particular importance of the latent variable measuring aspects of social capital. In addition, the path from the latent construct depicting risk and uncertainty the farmer faces to *inpa* becomes significant in the modified model at the 0.1 level. This was not the case in the initial model. While the other structural paths remain insignificant, the new significance of *unce* → *inpa* points to the fact that risk and uncertainty also play an important role with regard to program participation.

Having assessed the model parameters, the model as a whole has to be evaluated next. In the discussion about model estimation (see chapter 5.3) and inferential statistics (see Table 5-2) the χ^2 value was introduced as a rapid model appraisal tool. In addition, more practical model fit indices reported by AMOS® which are usually used to support the χ^2 statistic were introduced and discussed in chapter 5.4.

Table 5-19 presents the model fit summary for the modified model. As summarized in chapter 5.4 all fit statistics in AMOS® are reported for the three comparative models, namely the actually specified (or default) model, the saturated, and the independence model. Again, the interest is primarily in the default model results.

Table 5-19 Modified Model: Model Fit Summary: CMIN

Model	NPAR	CMIN	DF	P	CMIN/DF
Default model	64	141,382	126	,165	1,122
Saturated model	190	,000	0		
Independence model	19	1331,658	171	,000	7,787

For the initial model the chi-square value was 297,360 with 186 degrees of freedom and a probability level of .000. For the modified model, the chi-square value decreases to 141,382 with 126 degrees of freedom and a probability level of .165. How can the decrease of CMIN and DF be interpreted while the prob-

sistencies either. As discussed above, correlations greater than 1 and/or negative variances would have been a clear sign to discard the model.

ability level increases? CMIN in effect accounts for the discrepancy between *S*, the unrestricted sample covariance matrix and $\Sigma(\theta)$, the model implied covariance matrix. As outlined above it tests the extent to which all residuals in the residual matrix are zero. The smaller the chi-square value the better the actual model fit. However, chi-square has to be interpreted together with P, the probability value related to CMIN which stands for the likelihood to obtain a CMIN value that exceeds the actual CMIN value in the case H_0 is true. In 5.1 it was explained that the actual aim of SEM is not to reject the null hypothesis as it is intended that the postulated model is true in the population. Based on Table 5-19 H_0 for the modified model (see equation 5.1) cannot be rejected. The number of parameters (NPAR) in the modified model decreased as some observed variables from the initial model were excluded. Excluding variables means excluding data points (i.e., sample moments) available for model estimation which in turn decreased the degrees of freedom (DF) to 126.[68] The data points outweigh the parameters to be estimated in the modified model, the modified model is over-identified and has positive degrees of freedom. In addition, Table 5-19 reports the normed chi-square to lie in the desired range of values, thus, it can be concluded that the modified model fits the data not only substantially better than the initial model but has well-fitting properties per se.

In the following a slightly condensed subset of the various model fit indices discussed in chapter 5.4 is presented. The selected criteria were chosen based on their alternative approaches with regard to the assessment of model fit and their support in the literature (HOYLE (1995); BYRNE (2001)). As discussed above, for instance various model fit indices exist that address the issue of model parsimony, e.g., the AIC, the CAIC, and the ECVI. As those indices basically yield the same information, even though with slight modifications, it is always up to the researcher to decide which index to report in order to best inform about the fit of the respective model.

68 With 19 observed variables there are 19(19+1)/2, i.e., 190 data points available for parameter estimation in the modified model. The following parameters have to be estimated in the modified model: 19 regression parameters, 19 variances of the error terms, the variance of the residual term and seven factor variances. Finally, 15 factor covariances and three error term covariances (e4 and e5, e9 and e10, e7 and e8) which are not graphically represented in Fig. 5-4 have to be calculated, i.e., NPAR = 64. Data points for model analysis minus NPAR equal DF, i.e., 126.

Table 5-20 Modified Model: Alternative Fit Indices: RMR, GFI, AGFI, PGFI

Model	RMR	GFI	AGFI	PGFI
Default model	,268	,930	,895	,617
Saturated model	,000	1,000		
Independence model	,403	,573	,525	,515

Table 5-20 shows the alternative fit indices for the modified model: the root mean square residual (RMR), the goodness-of-fit (GFI) index together with its degrees of freedom adjusted counterpart, the AGFI, and the parsimony goodness-of-fit index (PGFI). The RMR represents the average residual value resulting out of fitting the model implied covariance matrix to the covariance matrix of the sample data. The statistic is shown here in its standardized form as average value across all standardized residuals with a range from zero to one. The better the model fits, the smaller the RMR value. Comparing the RMR of the modified model above with the RMR of the initial model (.280), a desired decrease of the statistic can be found. At the same time, the other fit indices have to be considered. The GFI as a measure for the relative amount of variance and covariance in S that is jointly explained by Σ and the AGFI, its DF-adjusted counterpart. The AGFI is one of the statistics addressing the issue of model parsimony. Both indices range from zero to one and the closer the value to one the better the actual model fit. The fourth fit index in Table 5-20 is the PGFI, which also addresses model complexity and penalizes more than the AGFI for the inclusion of additional parameters. As outlined before, the PGFI has a lower threshold value than the GFI and the AGFI. The PGFI of the initial model (.644) is slightly better than the PGFI for the modified version as in the modified model the number of sample moments decreased more rapidly than the parameters to be estimated. However, the GFI and the AGFI of the modified model indicate a much better fit than the values of the initial model (.876 for the GFI and .832 for the AGFI), which indicates a superior fit of the modified model. The GFI of .930 indicates that now 93% of the variance covariance in S is predicted by the model implied variance-covariance matrix. Next, the normed fit index (NFI), the Tucker-Lewis index (TLI), and the comparative fit index (CFI) contrast the proposed model with the independence model in order to evaluate model fit.

Table 5-21 Modified Model: Model Comparison: NFI, TLI, CFI

Model	NFI	TLI	CFI
Default model	,894	,982	,987
Saturated model	1,000		1,000
Independence model	,000	,000	,000

All three statistics are greater than the respective values for the initial model, which is a good indication for a superior model fit. The value range for all three statistics is from zero to one with higher values indicating better model fit. Formerly, values for both, the NFI as well as the CFI greater than .90 were considered to be indicative for a good model fit (BENTLER (1992)). However, as discussed above the NFI was reported to underestimate model fit especially in smaller samples (BYRNE (2001)). Nowadays the threshold for both statistics is .95 (HU and BENTLER (1999)). Given the sample size of 200 observations the relatively low NFI value of .894 for the modified model is not surprising. Still, this value is greater than the one for the initial model. However, the CFI now clearly meets the revised threshold of .95. Additionally, the Tucker-Lewis index (TLI) also meets the strict .95 threshold. The value range of this index is consistent with the other fit indices presented here. The parsimony-adjusted measures for the normed fit (PNFI) and the comparative fit index (PCFI), as well as the parsimony-ratio (PRATIO) are presented in Table 5-22.

Table 5-22 Modified Model: Parsimony-Adjusted Measures: PRATIO, PNFI, PCFI

Model	PRATIO	PNFI	PCFI
Default model	,737	,659	,727
Saturated model	,000	,000	,000
Independence model	1,000	,000	,000

All three indices also address the issue of model parsimony, which is the number of estimated parameters needed to achieve a desired level of model fit. The actual values of the three indices for the modified model are lower than those of the initial model. This detail is actually not unexpected given the fact that the ratio between distinct sample moments and distinct parameters to be estimated decreased for the modified model compared to the initial one. However, the values for all three indices give definitely no reason to discard the modified model. The next model fit index, the RMSEA, presented in Table 5-23 is one of the most informative and important indicators for model fit assessment (see chapter 5.4).

Table 5-23 Modified Model: Root Mean Square Error of Approximation: RMSEA

Model	RMSEA	LO 90	HI 90	PCLOSE
Default model	,025	,000	,041	,988
Independence model	,185	,176	,194	,000

Values less than .05 for the RMSEA represent a good model fit and it can be seen that the modified model easily meets this criterion. The 90% confidence interval reported around the RMSEA point estimate of .025 is excellent with the lower bound being .000 and the overall range of the interval being quite narrow. The test for the closeness of fit (PCLOSE) in the fifth column of Table 5-23 is also first-rate. It can be seen that the threshold value of .50 for the closeness of fit is easily fulfilled. All in all the root mean square error of approximation of the modified SEM is indicative of a very good model fit. It is obvious that the misspecifications of the initial model leading to higher RMSEA values were detected and corrected in the modified model version. The last fit statistic to be discussed here is the expected cross-validation index (ECVI).

Table 5-24 Modified Model: Expected Cross Validation Index: ECVI

Model	ECVI	LO 90	HI 90	MECVI
Default model	1.354	1.276	1.521	1.426
Saturated model	1.910	1.910	1.910	2.123
Independence model	6.883	6.318	7.485	6.904

As discussed, the ECVI is a measure to assess the likelihood that the hypothesized model for a single sample cross-validates in similar sized samples to be taken from the same population (BROWNE and CUDECK (1989)). There is no predetermined range of values for the ECVI. It is essentially a tool for model comparison. The model with the smallest ECVI has the greatest chance for replication. Comparison of the saturated and the independence model with the modified model ('default model' in Table 5-24) reveals that the modified model has the lowest ECVI of all three models. This means that it represents the best fitting model within the continuum of models between the saturated and the independence model. The confidence interval to assess the precision of the estimated ECVI value from 1.276 to 1.521 is very narrow around the 1.354 point estimate. In addition, it is narrower than the confidence interval of the independence model, which again indicates that the ECVI of the modified model is a very good approximation to the population.

In summarizing the model fit criteria for the modified SEM we can therefore conclude that all criteria are indicative of a very good model fit. Especially the root mean square error of approximation (RMSEA) of the modified model and the goodness-of-fit value should be stressed here again. Both values indicate an excellent model fit. In addition, the small cross validation index of the modified model with its very narrow confidence interval shows that it stands for the best fitting model within the aforementioned continuum of models. A second specification search did not result in any further recommended changes. Thus, the modified model is considered to be the best fitting structural equation model with regard to the sample variance-covariance data for the prediction of farmers' inclination to participate in the TIST program.

6. Synthesis – The Model Results and the Issue of Participation

In chapter 2, when discussing the issue of LULUCF projects as an option to mitigate carbon emissions, several potential prospects as well as pitfalls became already apparent with regard to the developmental benefits of such projects. The particular interest of the chapter was in factors determining the participation decisions of subsistence farmers in PES and more particularly land-use related CDM projects. Furthermore, chapter 2 raised concerns about impending obstacles for the participation of farmers in such projects. The aforementioned factors were the main focus of the modeling approach in chapter 5 given the fact that solely little knowledge exists with regard to the participation decisions of farmers in such programs up to this date (see PAGIOLA et al. (2005)). Most PES as well as carbon finance projects were identified as too recent for an assessment of actual participation decisions. Therefore, it was the aim of a broader literature survey to firstly find out factors determining participation in other rural programs. Subsequently, a theoretical economic model of adoption was developed for the specific circumstances of the tree-planting project studied during field research. Ultimately, the goal was to apply the findings of the reviewed literature as well as the theoretical model to the concrete case study in Tanzania by means of structural equation modeling. Looking back to the analysis of chapter 4, the most important determinants of adoption decisions described in the literature are compared to the model results of chapter 5.6 in the following.

In review of chapter 4, the quantitative results of PATTANAYAK et al. (2003) who analyzed an array of adoption studies with a vote-counting based meta analysis are of particular interest. PATTANAYAK et al. (2003) found preference proxies to be included in most of the adoption studies (i.e., in 47% of all the studies they reviewed), followed by factors assessing risk and uncertainty of the farmer's environment (39%) and the farm's resource endowments (38%). Market incentives (33%) and biophysical factors (27%) were less often found to be included in studies assessing factors of agroforestry adoption. With regard to significant exploratory contributions of the respective factors, PATTANAYAK et al. (2003) conclude that preference proxies, included in most of the studies, interestingly delivered least often significant results. The 'hotspots' with regard to significance were asset-related variables, variables measuring credit availability, price proxies, and the variable measuring contact to extension services. Variables measuring social capital (e.g., social connectedness) of the farm household were either not explicitly included in the studies reviewed by PATTANAYAK et al. (2003) or not included at all. In this regard the study at hand tried to shed light on a social aspect of adoption as will bc discussed in more detail further down. At first, the analysis proceeds with a review of the measurement and structural hypotheses established by the structural equation modeling approach, and verifies their supportability in the light of the modeling results.

It is important to keep in mind that according to SEM theory, all the path coefficients going from the latent variables on the left and right hand side of the model (*pref*, *rend*, *minc*, *biof*, *unce*, *pisc*, as well as *inpa*) to their respective indicators (*educ*, *age*, *gndr*, etc.) were all *measurement hypotheses* within the model. Given the fact that the whole SEM approach is a *confirmatory* one, it was *assumed* that the respective indicators are the best possible option available to explore the influence of hypothesized underlying latent variables. The idea of a measurement hypothesis refers to the fact that other unknown or untested proxies might also be suitable to depict the respective factor construct.

The relationships between the latent variables in the model establish six *structural* hypotheses. These hypotheses all describe the influence of the different latent variables on the left hand side of the model on the inclination to participate (*inpa*) on the right. The supportability of both types of hypotheses will be looked at in more detail in the following.

The comparison of the regression weights of the modified model in Table 5-18 with the theoretically established factor loadings in chapter 5.1 (see Table 5-1) reveals two interesting findings: (i) the respective signs in the model are mostly expected and consistent with the theory, and (ii) the huge majority of the observed variables loads statistically significant on the .05 level or better on their underlying constructs. In every case where the observed variable loads significantly on its underlying construct and in addition the respective sign is expected and theory consistent, the model validates the underlying measurement hypotheses.

For the majority of the measurement hypotheses of the model it can be concluded that the outlined indicators are very good proxies for their underlying latent constructs. Subsequently, a closer look is taken on the challenging relationships, i.e. those relations where underlying theory and model results are not so obvious at first sight.[69] The first noteworthy rather than contradictory result is the fact that *age* loads significantly negative on the latent preferences of the farmer. In chapter 3.3.1 it was discussed that the majority of household heads interviewed during field research was below 40 years of age. In addition, the age class distribution split into participants and non-participants revealed that the age of the household head amongst TIST participants was slightly higher than for non-participants. This finding reflected the 'snapshot' provided by the cross-sectional survey data. The analysis of chapter 5 however shows that when it comes to the actual participation decision, younger farmers with rather longer planning horizons seem to be more prone to participate. An explanation is the higher risk aversion of older farmers with regard to the introduction of a new technology as discussed in chapter 4.1. The opposite effect discussed above, namely that a higher age of the farmer is linked to greater farming experience, a

69 Only *free* parameters that are unknown and therefore subject to the estimation process can be interpreted in more detail. 'Reference variables', i.e., measured variables with a fixed factor loading to determine the scale of the construct have a pre-specified value and are therefore not subject to estimation and interpretation.

greater accumulation of knowledge and thus a greater openness with regard to the introduction of new technologies must be negated in the actual context.

In the case of off-farm income (*incm*) a household realizes through any kind of wage earnings in other business activities, the model of Fig. 5-4 suggests that a higher off-farm income is a significant proxy for greater market incentives, which is plausible. Furthermore, the correlation of off-farm income with adoption was found to be mostly negative in the literature review of chapter 4.1. The reasoning was that no interest to adopt a new and potentially risk-prone technology might exist where off-farm income already satisfies family needs (KEBEDE (1992)). This finding described in the literature corresponds properly to the negative structural path going from market incentives (*minc*) to the farmer's inclination to participate in the program (*inpa*). Thus, the positive factor loading from *incm* on *minc* for greater market incentives and the negative loading from *minc* on *inpa* are no contradiction but an interesting model feature that reflects the underlying theory. A long-term and potentially risky technology like tree planting is not so attractive where off-farm income sources already fulfill requirements to nourish a family.

Another interesting relation exists in the case of the latent bio-physical farm characteristics. In Table 5-1 it was assumed that a better soil quality of the farm is rather negatively correlated with a tree planting decision as farmers are supposed to use better sites for food production and rather plant trees on marginal lands. In addition, it was hypothesized that greater land holdings make the investment in a new technology more feasible. From the model results in Table 5-18 it can be seen that both soil quality and farm size load positive and significant on the latent variable depicting the biophysical farm characteristics. Thus, the signs in the measurement model make perfectly sense. In the structural model between *biof* and *inpa* however, a negative sign becomes obvious which is contradicting the underlying theory at first view. However, the relation reveals that farmers with better biophysical farm characteristics rather abstain from program participation. In reverse, it could be argued that the model indicates that resource-poorer farmers have a greater inclination to participate than their better-endowed colleagues. Yet, this should be taken with caution because the interrelation is not significant so that it is difficult to come to an appropriate interpretation of the result. But still it can be argued that this effect is an interesting feature of the case study even though it is rather weakly pronounced.

Furthermore, the measurement model of *inpa* reveals another interesting aspect. The factor loading from the gender variable on the latent inclination to participate is negative in the model. A finding that contradicts the respective measurement hypothesis established above in Table 5-1. The gender variable measures the proportion of male members in the household and it was assumed that a higher male household member proportion would rather positively influence the inclination to participate in the program. The main reasoning for this relationship is the higher risk-tolerance observed for male household members and a greater labor availability which in turn positively influences the adoption decision. However, the contrary is observable for the data structure of the structural equation model. Obviously, more male household members have a rather negative influence on the participation decision. This model result is consistent with the observed gender distribution of the interviewed farm households. In chapter 3.1.1 it was shown that significantly more female-headed households are found within the group of TIST farmers than within non-TIST farm households. About 10% more female-headed households were found amongst TIST members. While above caution was expressed with regard to the relationship between the actual participation decision and the gender aspect of the household, the model results reveal that there is a significant relationship between a female-headed household and the household's inclination to participate in the program.

The signs of all other factor loadings in the measurement models are expected and theory consistent. In addition, they are all significant which again justifies their explanatory power in the measurement models. For the reason not to replicate too much, they are discussed here rather briefly. The free parameter value in the measurement model of the household's preferences was already presented above (see discussion about *age* influencing the farmer's participation decision). The second measurement model depicting the resources of the farm shows that household assets contribute significantly to a higher resource endowment of the farm household, which is also theoretically valid. The same is true for the sum of the household's expenditures, which is positively correlated to the household's resource endowments. The latent variable *rend* in turn has a positive impact on the household's participation decision, even though this relationship is not significant, it is theory consistent.

The relationship between off-farm income (*incm*), the latent market incentives (*minc*) and the farm household's inclination to participate in the program was also discussed above. Two more aspects shall be added here, namely that expectedly the price proxy (*prce*) loads negatively on market incentives in the measurement model, i.e., the higher market prices the lower are market incentives for the farm household. In addition, the relationship between market incentives and *inpa* is negative but not significant in the structural model. We will come back to the issue of significance in the structural models in the following.

The essential aspects with regard to the measurement model of the biophysical farm characteristics have been discussed above so that we proceed to the measurement model for risk and uncertainty. Within this measurement model a higher proportion of plots with restricted access (i.e., the proportional tenure restriction measure, *rest*) loads statistically significant at the 0.1 level on the latent variable depicting risks and uncertainties in the farming environment. This finding was expected in Table 5-1. However, the significantly positive relationship (at the 0.1 level) in the structural model between these farm environment uncertainties and the household's inclination to participate is of particular interest because it clearly contradicts the structural hypothesis in Table 5-1. The literature review on agroforestry adoption in chapter 4.1 highlighted that the household's attitudes towards short- and long-term risks and uncertainties in the farm environment determine production decisions related to resource allocation and technology choice. In addition, several authors found that in particular poverty and economic risk reduce the incentive to invest in long term farming activities like tree planting (SCHERR (1995a) REARDON and VOSTI (1995)). The model results however indicate that higher uncertainties in the farm environment *positively* influence the household's inclination to participate. Thus, farm households perceive participation in the TIST program as a risk mitigation option. One reason might be the expected and contractually set down carbon finance payments that provide a continuous stream of revenue, which helps to smooth farm income fluctuations. As discussed earlier, the new and innovative characteristic of carbon finance payments for sequestration services is that they substantially reduce the long gestation period of a tree planting activity and provide returns even prior to maturity of the trees for other purposes (e.g., timber, fruits, medicine etc.).

Besides the aforementioned interesting aspect with regard to the gender proxy in the measurement model for the farm household's inclination to participate, the proxy for the religious affiliation of the farm household shows the hypothesized characteristics. Anglican denomination of the household head loads positively on the household's inclination to participate in the program. Given the fact that the Anglican Church initially promoted the program this is a rather likely finding. However, one of the main objectives of the TIST program is to broaden its outreach in order to give all interested farm households, regardless of their denomination or religious affiliation, a chance to participate. Given the model results it is worth to look at the targeting of the program in more detail in the future in order not to narrow the chance for program participation a priori by only addressing specific groups of the society, e.g., a specific denomination.

The last measurement and structural relationships to be looked at are those between the 'perceived indicator of social capital' (*inpa*) and its measures as well as between *inpa* and the farmer's inclination to participate in the program (*inpa*). In chapter 4 it was shown that variables measuring social capital and specifically its impact on adoption were hardly included in any of the reviewed studies. Therefore, it was the aim to shed light on some newly hypothesized relationships between social capital and adoption. As outlined above, the newly introduced concept of the structural equation model presented in chapter 5, is the latent variable describing social interactions of the rural farm households and their impacts on adoption. For this purpose, a perceived indicator of social capital is measured in the modified structural equation model of Fig. 5-4 by the three variables *memb*, *n_ext*, and *vsts*. While *memb* counts memberships in different community organizations other than the TIST program, *n_ext* reports the quantity of extension contacts in the last twelve months and *vsts* highlights the actual visits per year of community or church group meetings. All three were hypothesized to measure the 'connectedness' of the farm household in the community based on self-reported data. While *vsts* is the reference variable to determine the scale of the latent *pisc*, Table 5-18 shows memberships as well as the variable measuring extension contacts to load significantly (at the .001 level) on *pisc*.

Most importantly, the latent variable *pisc* has the strongest impact (significance at the .05 level) on the farm household's inclination to participate in the program. The initial model demonstrated the same effect (see Table 5-3), which underlines the particular importance of the latent variable measuring aspects of social capital. Through social interactions, the farmers are able to discuss on the one hand helpful, on the other hand disappointing experiences, circulate information, and pass on precautions with regard to a prospective technology. As in the case of farm environment uncertainties, which were mitigated through program participation, it can be concluded that better-connected households have a higher inclination to participate in the program, as they are more capable to adapt to potential risks of the new technology due to more information sources. Thus, the higher the perceived indicator of social capital the better the information network of the farm household, which mitigates risks and uncertainties related to the adoption of a the new tree planting technology.

To summarize the model results with regard to participation: the significant results of the measurement models indicate that the hypothesized factors underlying the observed and measured variables are reasonable and acceptable. In the structural models the respective signs in the regression structure among the latent variables *pref* and *inpa*, *rend* and *inpa*, and *minc* and *inpa* are all theory-consistent. The structural relationship between the bio-physical farm characteristics and the household's inclination to participate in the program was found to contradict underlying theory at first view. However, it was concluded that farmers with better bio-physical farm characteristics rather abstain from program participation. Resource-poorer farmers have in principle a greater inclination to participate than their better-endowed colleagues. These results have to be interpreted with caution, as the model specific impact of all the aforementioned latent variables on *inpa* is rather weak (i.e., insignificant).

But two significant structural relationships exist between *unce* and *inpa* and *pisc* and *inpa*. Those two relationships are of greater value with regard to a meaningful interpretation than the insignificant ones, which are fairly weak. However, what do the weak relationships for household preferences, resource endowments, market incentives and bio-physical farm characteristics actually mean? They indicate that those factors are not of such great importance for the case at hand. Going back to the PATTANAYAK et al. (2003) meta-study, household assets were for example found to be significant at almost all times. Thus, the PATTANAYAK et al. (2003) review identified them as a major driver for a positive adoption decision. In our case, however, given the insignificant regression paths in the modified SEM (see Table 5-18) it is reasonable to assume that this finding cannot be extrapolated for the reforestation program under TIST.

6.1. Asset-based Project Impacts on Participants

In chapter 2.4 on carbon mitigation and development benefits of LULUCF projects, the focus was on the one hand on general considerations with regard to program design. On the other hand, three main questions with regard to the interlinkages of environmental service payments and poverty were highlighted. They were about actual and potential participants of PES or carbon finance schemes, and factors influencing their participation decision. In addition, the actual impact of the payment scheme on participants was considered to be important (PAGIOLA et al. (2005)). Having assessed aspects addressing the first two questions, i.e., participation issues in the preceding chapter, the focus of chapter 6.1 is a rather qualitative one by looking on the so called asset-based project impacts on farmers participating in the TIST project. To this end, we return to the 'Sustainable Livelihoods Approach' (SLA) introduced in chapter 2.4. CHAMBERS and CONWAY (1991) present the SLA as a way to foster the general understanding of the livelihoods on which poor people depend. In doing so, a SLA draws on the main factors that affect the livelihoods of poor people and outlines the typical relationships between these factors. SLAs are suitable on the one hand to plan new development activities and on the other to assess the contribution of existing activities to sustaining livelihoods. The latter aspect is of main importance when it comes to the assessment of the CDM twin objective with regard to climate protection and fostering sustainable development. The CDM framework makes clear that the CDM host country has the sole prerogative to define 'sustainable development' under any CDM project category according to its specific national circumstances. Nevertheless, the SLA provides a good common framework to assess CDM forestry projects with regard to their component fostering sustainable rural development. This particularly includes an assessment of forestry projects regarding their potential to contribute to rural poverty alleviation.

'Rural poverty' is usually understood not only as description of the lack of goods and services to fulfill ones daily needs but includes other dimensions as well. These other aspects can be addressed through different indicators, e.g., level of income and consumption, but also social indicators, and indicators of vulnerability to risks as well as indicators of social and political access (THE WORLD BANK (2005)). The interest of the SLA in this context is to address and understand the sustainability contribution of an activity from a rather holistic point of view. The SLA evaluation describes how the activity impacts on the broader asset base on which project participants depend. To this end, the six asset classes introduced in Table 2-2 (i.e., natural, social, human, physical, financial, and political assets) are revisited. While Table 2-2 has merely set out opportunities of PES or carbon finance markets against risks that potentially devalue those key assets, Table 6-1 adds a third column, which presents key features of the TIST program within those asset classes. Potential risks as well as additional opportunities are further elaborated.

6.1.1. Natural Assets

An opportunity of carbon finance payments for afforestation is the increasing value of formerly marginal lands due to the new market opportunities. In addition, positive spill-over effects for timber and non-timber forest products (NTFPs) have been reported in the literature when sustainable forest exploitation is permitted on top of the delivery of the environmental service (LANDELL-MILLS (2002)). However, opportunity costs arise where the delivery of the environmental service entirely prohibits or hinders the production of timber and/or NTFPs. Furthermore, there are positive impacts on other related natural assets. Soil fertility and agricultural yields increase due to the prevention of erosion. The reduction of erosion also ameliorates water flows. Moreover, afforestation and reforestation activities have the evident opportunity to improve local air quality.

Potential drawbacks and risks in the area of natural assets are that the new markets might be inaccessible by poorer farmers due to several reasons. Uncertain property rights, insufficient market knowledge, information, and missing marketing skills due to lacking education are amongst them. Further reasons are the complete lack of contracts or inappropriate contract design like rather long-term contracts that might be unsuitable for rural livelihood strategies that often require rather short-term flexibility. How does TIST deal with these aforementioned opportunities and risks? To start with the latter issue of contract design: TIST enters into specific GHG sale agreements with established Small Groups. The Small Group (i.e., 5 to 12 farmers) forms the unit of aggregation to address the issue of transaction costs. The TIST sponsor, Clean Air Action Corporation (CAAC), intends to enter in an agreement with the entire Small Group not with the single farmer. The agreement between CAAC and the Small Group is valid for up to fifty years. However, the individual farmers have to commit themselves only to participate in the respective TIST Small Group for at least five years. This contractual design must apparently handle the issue of farmer 'drop-outs' from the Small Group. However, no specific solution addressing possible drop-outs was observed during field research. This ultimately raises the question of the program's long-term continuity and therewith the permanence of the service provision itself.

While the literature often rather addresses the issue of appropriate contract length, in order not to 'lock' farmers into a specific land use form, it is important to consider more flexible contractual solutions. In addition, it is useful to differentiate between the contract scheme and the actual payment scheme as will be explained further down. Subsistence needs of rural farm households determine the farmer's planning horizon to be rather short. Field research results reaffirmed this general finding, as almost one third (31.6%) of all farmers preferred a shorter contract length of only five years. Only one per cent of all farm households were convinced that a 50-year contract is actually appropriate for their livelihood situation. However, when given the theoretic opportunity to renew the

GHG sale agreement in five-yearly intervals, more than two-thirds (67.3%) of all households would be willing to accept a contract length of up to 60 years. This finding is worthy of note given the internationally agreed flexibility to issue temporary certified emission reduction units (tCERs) for CDM forestry projects as described in chapter 2.2. Thus, tCERs not only address the issue of potential non-permanence of forestry projects but also offer a suitable livelihood strategy for rural farm households. With regard to the payment scheme for carbon sequestration services, 50% of the farmers prefer to be paid every third month and 16% would even prefer a shorter period of regularly payments (e.g., one month). Only 14% of the farmers consider extended payment periods for their service provision as long as one year. This underlines the regular income effect sought by afforestation activities. Such frequent payment intervals are clearly a challenge for program sponsors due to the remoteness of project sites and participating farm households. However, the actual design of the payment scheme provides an opportunity to align rather long-term contractual solutions preferred by the program proponent with the livelihood needs of rural farmers.

As already outlined above, an emerging market for environmental services can be inaccessible for poorer farmers due to uncertain or lacking property rights. The issue of land ownership and land tenure is a serious one because in principle only land-owning farmers can participate in the market. GRIEG-GRAN et al. (2005) found landless people often to be excluded by the specific entry regulations of PES schemes, if either a minimum land size or unambiguous land titles are an access eligibility requirement. The challenge when it comes to the assessment of intended program impacts on peasants is the fact that often the poorest are landless. This kind of 'landless farmer' may only own some very few hectares or acres of agricultural land, which cannot be set aside for the CDM-motivated afforestation activity. Clearly, food production for subsistence of the family is the major occupation of the household. Thus, in principle this type of farmer is excluded from service provision. Yet, the results of the former chapter show resource-poorer farmers to have a greater inclination to participate in the TIST program than their better-endowed colleagues do. TIST actually encourages farmers with rather pronounced land constraints to participate in the program. Through the described Small Group structure, trees can even be plan-ted collectively on non-owned farm sites while the carbon finance revenues are shared within the group. Furthermore, PAGIOLA et al. (2005) rightly pointed to the fact that the lack of titles should not be set equal with tenure insecurity because titles in developing countries are often rare and may not be necessary as long as tenure is secure.

Figures from the field survey reflect this finding largely. The interviewed farm households had only for 3.1% of all the plots they cultivated an official certificate of ownership while as much as 79% of the plots were farmed 'only' under customary law, i.e., without the existence of an explicit title. However, as reported in chapter 3.3.5 only 22.4% of non-TIST and just 7.8% of TIST farmers understand this fact as an actual restriction of their tenure rights. Farming takes place in most of the cases on customarily 'owned' fields. Almost always farmers are aware of the borders between different fields and farms and the TIST program encourages them to mark those borders by planting trees around homes, farms, roads, and alongside paths. Thus, TIST takes a hands-on approach to promote the formalization and clarification of resource tenure rights. However, lacking access control over and access restriction to areas where TIST trees have been planted was found to be a major problem. As outlined in chapter 2.4 missing access control is an unreliable precondition for carbon sequestration service provision. In the TIST program, livestock encroachment and grazing were clearly found to be a problem in several of the afforestation areas.

Sustainable forest exploitation is not only permitted by the TIST program but also actively encouraged. It clearly provides the opportunity for positive spill-over effects between carbon sequestration service provision and the enhancement of timber and non-timber forest products and services available to rural farmers. Positive impacts on other natural assets, i.e., the fertility of the soil and hence agricultural yields can be studied: the tree types planted by TIST farmers prevent erosion and thus improve the soil quality. Amongst the most frequently planted tree species are *Leucaena leucocephala* (Lam.) de Wit, *Senna siamea* (Lam.) H.S. Irwin & Barneby, and *Moringa oleifera* (Lam.). *Leucaena leucocephala* is most suitable to restore forest cover in degraded areas as it grows very well on marginal sites with extended dry seasons and on steep slopes (MBUYA et al. (1994)). With its high nitrogen-fixing potential of up to 100-300 kg N/ha a year, related to its abundant root nodulation, *Leucaena leucocephala* restores soil fertility. Its pronounced taproot system is able to break up compacted subsoil layers. This improves the infiltration of moisture into the soil, reduces surface runoff and hence helps to control erosion (ICRAF (2006)). *Senna siamea* is outstanding with regard to its biomass production. A well-grown tree can yield 500 kg/year of fresh leaves, which are used as green manure and fodder. By contrast, the leaves of *Moringa oleifera* are mainly used for human food and only seldom for fodder. The leaves enrich the farmers' diet with protein, and the vitamins A, B and C. In addition, minerals such as calcium and iron are contained in *Moringa oleifera.* YANG et al. (2006) report the tree leaves to be an excellent source of the sulphur-containing amino acids methionine and cystine, which are often in short supply. Young pods and peas as well as the tree flowers are edible and the seeds are either eaten or used to produce very tasty oil (ICRAF (2006)). With regard to their fuel and timber value, the three species differ markedly. While *Leucaena leucocephala* is a very suitable firewood species, which produces only very little smoke, almost no flying sparks, and has a high calorific value, *Senna siamea* and *Moringa oleifera* produce more smoke, which makes them less suitable for domestic fuel. Indoor air pollution can cause respiratory health problems and is a serious health issue to address (see further down: human assets). The charcoal of *Leucaena leucocephala* and *Senna siamea* is of very good quality and often sold on local markets, which generates additional income. By contrast, *Moringa oleifera* has a soft and weightless wood, which is almost unsuitable as firewood for cooking. The timber is useful only for light construction work. In addition, only very poor charcoal is produced from *Moinga oleifera* (FUWAPE and AKINDELE (1997); KUNTASHULA and MAFONGOYA (2005)). In contrary, both *Leucaena leucocephala* and *Senna siamea* produce a rather hard wood, used for a great variety of carpentry purposes. While *Leucaena leucocephala* is easily workable, it is a greater effort to use the durable, dark heartwood of *Senna siamea* for joinery. However, the latter is resistant to termites while the former is rather limited in use because of the lack of durability, and its weak resistance against termites and woodborers (ICRAF (2006)).

Overall, the three major tree species planted by TIST farmers in Tanzania make up an excellent portfolio, which comprises many livelihood-relevant timber and NTFP uses. The tree portfolio is completed by a dozen other native and non-native species planted by TIST farmers. Amongst them *Adansonia digitata* L., which provides an array of products and services to the rural household. The leaves are a medicine to treat bladder and kidney diseases, asthma, diarrhea, insect bites, parasitic worm infections (e.g., the guinea worm) and general fatigue. Antihistamine properties are not only present in the leaves but also in the flowers, which are valued highly amongst farmers as infusion for respiratory problems and digestive disorders (ICRAF (2006)). Many other important uses are reported in the literature, e.g., a paste of the seeds is used for curing tooth and gum diseases or gum from the bark, which is used for cleansing sores (HINES and ECKMAN (1993)). For apiculture *Adansonia digitata* is a source of fine quality honey. Wild bees perforate the soft wood and depose their honey in the holes. In Tanzania as in many parts of Africa, the hollow trunks are used for beekeeping. With regard to fruits, the mango tree (*Mangifera indica* L.) is highly appreciated by rural farmers as it enriches the diet with vitamins A and C. Furthermore, its umbrella-shaped crown makes it a suitable shade tree for people and livestock. It provides windbreak and improves the microclimate around human settlements. For planting the aforementioned trees, TIST participants gather their own seed locally and in some cases swap seed with other Small Groups in order to avoid monoculture planting and genetic bottlenecks. In general, the program does not import any seeds.

Besides the tree planting component, TIST encourages the farmers to shift their production to conservation farming techniques, which include minimum tillage systems and the practice of planting the trees in holes to reduce water run-off (FAO (1997)). The digging of planting holes concentrates rainfall around the seeds, which in turn accelerates emergence of the crop and improves the crop stands. In addition, minimum tillage systems, i.e., retaining residues on the fields also reduces soil and water loss, improves the infiltration of the soil, lowers surface temperatures, and ultimately once more improves soil fertility. TIST is training Small Groups in conservation farming practice. Up to this date, approximately 600 hectares have been planted using conservation farming techniques. Other agricultural techniques like crop rotation or terracing are discussed by the Small Groups and implemented where feasible. All things considered, conservation farming minimizes crop loss in drought years and improves food security (FAO (1997)). Thus, conservation farming not only has a positive impact on the natural assets on which local farmers depend but is also an additional feature of TIST to improve farmers' livelihood situation.

6.1.2. Social Assets

With regard to opportunities in the social asset class, Table 6-1 outlines that an upcoming market for environmental services may foster the formalization of resource tenure rights and clarify property rights over environmental services. Community-based organizations are encouraged to collaborate in the delivery of the environmental service, which leads to increased organizational and management capacity. However, increased competition for control over land in order to be able to provide the remunerative service may ultimately lead to social tensions, which would be an undesired outcome of carbon finance payments. The splitting up of communities in households gaining and households losing due to the market is a potential threat to local culture, as environmental services become commercial assets. TIST responds to those concerns with the voluntary participation in TIST Small Groups. Small Groups are open for everybody and all farmers have equal rights in the Small Groups. Every Small Group has a 'spokesman' and the leadership of the Small Group is supposed to rotate at least monthly. The rotating leadership is also a program feature trying to prevent potential rent seeking of the more powerful farmers within the group.

The field survey gathered information about potential rent seeking by asking how many different leaders the interviewed Small Group already had since its formation. The assumption was that, as the number of different leaders increases, the respective Small Group is less prone to rent seeking compared to other Small Groups. The survey revealed that 44% of all Small Groups had only one leader since the formation of the group. An almost equal percentage of Small Groups had no more than two (27%), or respectively three (25%) different leaders since group formation. All interviewed Small Groups existed for more than two years at the point the interview was conducted. Finally, none of the Small Groups had more than four leaders since its establishment, which is surprising given the fact that TIST requires the leadership to alternate on a weekly basis as mentioned before. However, asking for the frequency of Small Group meetings reveals that 52% of all Small Groups meet only once or twice a month and 19% do not even meet monthly but rather irregularly, e.g. every six or seven weeks. Only a fraction of 18% meets four times a month as required by the regulations of the TIST program (TIST (2004)).

As noted above, TIST Small Groups develop 'best practices' during their planting activities, e.g., for establishing nurseries and caring for the seedlings, for germinating and transplanting different tree species or for the selection of different tree species. These best practices are disseminated amongst TIST groups through a regular newsletter delivered to Small Groups, written in both English and Swahili.[70] In addition, the Small Groups meet on a regular basis in so called 'community node meetings'. At those meetings, several groups living in the same area get together and share their experiences. In addition, a TIST responsible person (i.e., a 'TIST coordinator' coming from the TIST head office in Morogoro, Tanzania) collects the 'Small Group Monthly Reports' about the actual number of trees planted and maintained.[71] The monthly report is the basis for the yearly program-internal auditing process. Thus, a number of approaches are set up by the program to foster the collaboration of communities and Small Groups. A problematic split in terms of economic disparity of households due to the new market opportunities is not observable. However, the actual working of Small Groups in the field does not meet the standards outlined by TIST program officials. The educational aim of the Small Group meetings directly addresses human assets as well. Human assets relate to the educational and skill base of the farmers participating in the program.

6.1.3. Human Assets

Not only the TIST newsletter, also the Small Group and the community node meetings lead to an improved education and skill base relating to forest management amongst TIST participants. Almost 60% of Small Group members felt that their knowledge improved in the aforementioned field due to the participation in TIST. Referring to Table 6-1 it becomes obvious that TIST is able to foster human assets in several different ways. Besides strengthening the cooperation amongst TIST farmers, the project also addresses health issues. This is done indirectly through an improved diet due to the specific species planted (as outlined above) and directly by addressing health issues like HIV/AIDS and malaria. In addition, issues related to childcare, and nutrition are discussed during Small Group meetings. Furthermore, the Groups report to develop orphan care strategies (TIST (2003)). Thus, the carbon sequestration activity links to a broader context of issues concerning everyday life.

In spite of this, it remains doubtful if farmers are fully aware of the broader climate-relevant context of the tree planting activity (i.e., carbon sequestration for Kyoto Protocol commitments, CDM transactions, etc.). Only 14% of TIST farmers have ever heard of the Kyoto Protocol. When asked for their mo-

---

70 See the newsletter section on TIST's web page to get the actual newsletter http://www.tist.org/moreinfo.php#tanzania

71 Small Groups who have handed in three consecutive Small Group Monthly Reports are eligible to receive their carbon finance vouchers.

tivation, the two most frequent responses explaining program participation are carbon finance payments (27% of all TIST farmers) and TIST being a source of cheaper firewood (also 27%). It is therefore unlikely that carbon finance payments are actually put into a broader environmental context by the farmers.

One of the concerns raised with regard to poorer households being excluded from service provision (see Table 6-1, Human Assets, Risks) is that those households might only be able to work in subordinate jobs. Meanwhile indispensable skills for the service provision are brought in externally. However, this cannot be upheld for the TIST program. The Small Group approach is integrative and open for farmers with different educational and wealth backgrounds. This was confirmed by the results of chapter 5.6 and 6. Farmers keen to participate more actively in the program can even get further education and become 'Quantifiers' who travel to various locations of the program and record group results (i.e., number of trees planted, diameter, tree height, tree species, age of the planting etc.). To do so, they attend training on hand-held computers and Global Position System (GPS) technology to exactly localize the tree groves. In addition, Quantifiers are taught basic features of Internet communication to transmit data. In this manner, TIST creates a communication and administrative structure by investing in improved communication links to remote forest areas.

6.1.4. *Physical Assets*

The aforementioned organizational structure is a quasi-physical asset, which connects rural livelihoods to the 'outside world'. Via the communication established by the project not only program-related information can be exchanged but an array of important information can be gathered and disseminated more rapidly, e.g., market prices for agricultural goods, employment opportunities in a nearby village etc.

6.1.5. *Financial Assets*

Regarding the project impacts on financial assets of TIST members, the survey firstly shed light on the question how many farmers actually received payments for their tree planting activities. In theory, carbon finance payments provide a diversified income base, as discussed. The payments should be a regular revenue with the prospect to be paid over many years. However, in practice the remoteness and scatteredness of the participants is a serious problem when it comes to the distribution of the payments on the ground. TIST does not directly deliver the carbon finance payments to the Small Groups but distributes carbon finance vouchers for the tree planting activity. Only Small Groups with more than 1,000 trees alive and a regular reporting to TIST about their activities are eligible to receive a voucher. The Small Group can then deposit the voucher at a local bank. Thus, in order to be able to cash the voucher and receive carbon finance money, the Small Group must set up a bank account.

Overall, there are three major pre-requisites to be fulfilled before carbon finance payments eventually become effective: (i) the program-specific requirements have to be met (i.e., 1,000 trees alive per Small Group and regular reporting), (ii) a voucher has to be received and (iii) a bank account has to be opened. How this affects the overall outcome of the payment scheme portrays Fig. 6-1. Amongst all TIST farmers, solely 37% actually received a voucher for their activities *and* cashed this voucher at a local bank where their Small Group had established a bank account. Only this fraction of TIST farmers has actually made income out of their tree planting activities. In contrary, 36% of the TIST farmers were not eligible for a voucher and hence not eligible for payment. The ineligibility results from the fact that the Small Group fell either below the '1,000-alive-trees' criterion (53% of the ineligible Small Groups) or the Group had not submitted three consecutive monthly reports and therefore did not qualify to receive any payments from TIST (see TIST (2004)). The remainder of TIST farmers (27%) is in principle eligible to receive carbon finance payments according to program requirements, however, due to two major reasons the payment scheme as designed by the TIST program fails to deliver the reward. On the one hand, it becomes clear that the distribution of TIST vouchers is unsuccessful. 18% of the interviewed farmers were eligible according to program requirements but did not receive any voucher. Not only the program coordinators of TIST are responsible for this malfunction, but there is another important point to highlight. The opportunity costs of transportation in the survey area are rather high so that TIST farmers sometimes do not attend Small Group and community node meetings where they would receive a voucher. Nine percent of TIST farmers were eligible and had received a voucher but were not able to cash the money because they had no bank account established. Again, costs of transport to the village or rural town with a rural bank branch and possibly additional costs to open the bank account are higher than the expected TIST carbon finance payments.

Fig. 6-1 Overview: Effectiveness of TIST's Carbon Finance Payment Scheme

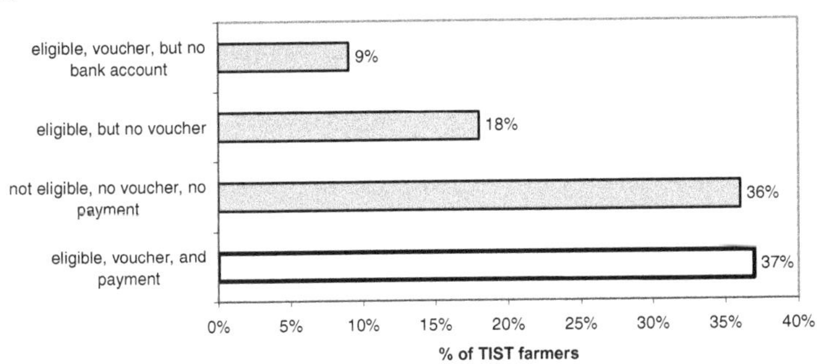

Thus, it becomes clear that (i) the voucher distribution and (ii) the costs for opening a bank account are major bottlenecks with regard to a successful carbon finance transaction apart from the eligibility requirements.

Given the fact that eight out of ten of the *eligible* farmers *receiving* a voucher are also able to cash it (payment success, see Fig. 6-1), the payment success rate could be ceteris paribus increased from currently 37% to about 51% if all eligible farmers would receive a voucher. That is, if the TIST program would optimize the voucher distribution and only the second bottleneck remains. The second bottleneck is the transaction costs imposed on TIST participants by opening a bank account. The rural banking system in Tanzania is clearly not the focus of the study at hand, however, if more efficient money channels could be found, and program officials would be willing to take at least part of the transaction costs for the bank account opening, the payment success could be substantially increased (up to 64% success rate).

As the respective farmers do not receive the carbon finance money directly themselves, but rather the Small Group they are belonging to, the distributional regime of the Small Groups also has to be considered. The survey revealed that 87% of the interviewed farmers (belonging to 34 different Small Groups) distributed the money equally amongst all group members. Only 13% of the interviewed reported other distributional rules. Here, the 'number of trees planted by farmer' comes second. However, empirical research in the field did not show that Small Groups have a formalized system to track plantings by the respective group members. This fact also casts doubt on other reported distributional scheme, like labor contributed to the group's planting activities.

Out of the 37% of TIST member farmers that received a voucher, 81% actually distributed the money among their group members. Interestingly enough 19% did not allocate the TIST money within their respective Small Group but rather preferred to leave it on the bank account. The major reason for that is the fact that carbon finance money is perceived as a kind of reserve for times of hardship or crop failure due to drought etc.

With regard to the remoteness of the participating farm households and its potential effect on voucher distribution and thus payment success rates, we once again revert to the map of the survey area in central Tanzania (see chapter 3.2.2). The question is where do the farmers receiving payments actually reside? Is there an indication that those farmers are somehow clustered or living near transportation routes or villages that are better accessible? As mentioned earlier, the survey area covered two regions (Dodoma and Morogoro) and three districts (Mpwapwa, Kongwa, and Kilosa). All the farmers that did receive a payment for their activity live in the region of Dodoma and here in the districts of Mpwapwa and Kongwa. Interestingly enough farmers being paid live in Gulwe, Bumila, and Mima, which all lie on the same road going from the district capital of Mpwapwa to the Rubeho Wota Plateau and then further south to the regional capital of Iringa. The villages of Chamae and Ngomai where payments are also successful both lie approx. 15 km north of the tarred principal highway connecting Dar es Salaam with Dodoma. The other two villages (Mlali and Manghweta) are located on the road connecting the district capital of Mpwapwa with the principal highway to Morogoro via Gairo. Thus, the proximity of a suitable infrastructure is a major success factor for the reward of the service.

6.1.6. Political Assets

As noted above, the characterization of 'rural poverty' aims at a multi-dimensional definition of poverty. Not only the absence of resources to fulfill daily needs is comprised under this definition but further facets. A prominent role play aspects of social and political access (THE WORLD BANK (2005)). The lack not only to influence political decisions but also the nonexistence of channels to articulate necessities, to make oneself heard, is a crucial characteristic of poverty. Table 6-1 shows that a potential positive impact of TIST's carbon finance program on the political assets of participating households is that the formation of community-based organizations provides the crucial basis for gaining political representation and voice. As discussed earlier there are risks involved, too. If the emerging of markets for environmental services like carbon sequestration leads to the marginalization of poorer farmers (e.g., through access restrictions because of missing land titles etc.), these excluded peasants have even less channels to influence political decisions. TIST addresses this situation by requesting farmers to actively contribute to the continuous program development. The Small Groups are encouraged to submit not only a monthly report about their tree-planting activities but also suggestions for best practices and program improvements. The submitting Group agrees upon those best practices in advance, thus, the local democratic decision making of the Small Groups is strengthened. Depending on how responsive the local TIST office in Morogoro, Tanzania and the program sponsor CAAC are to those submissions, the Small Groups might be able to relate a changing program design to their own suggestions.

Table 6-1 Potential Impacts of TIST Carbon Finance Program on Assets of Participating Households

Opportunities	Risks	TIST Approach
	Natural Assets	
Increased value of marginal lands due to new market opportunities.	New markets inaccessible by poorer farmers due to uncertain property rights, insufficient marketing skills and education inadequate information, lack of contracts, inadequate communication structure, inappropriate commodity design, e.g., long-term contracts unsuitable for livelihood strategies that require short-term flexibility.	Trees used to mark borders around homes, farms, roads, and paths. Greenhouse gas sale agreements with Small Groups existing. Contract duration up to fifty years.
Positive spill-over effects for timber and non-timber forest products (NTFPs) where sustainable forest exploitation is permitted.		Sustainable forest exploitation permitted. Provision of fruits, nuts, fodder, timber, medicines, bee habitats (i.e., honey). Groves of trees as village woodlots. Introduction of sustainable agricultural techniques.
Positive impacts on other related natural assets (i) soil fertility and agricultural yields (ii) water flows and quality (iii) air quality	Opportunity costs where delivery of the environmental service prohibits / hinders production of timber and / or NTFPs.	Trees for erosion prevention. Specific tree types improve soil quality. Provision of shade and windbreaks. Micro-climate improvement.
	Social Assets	
Market may foster the formalization of resource tenure and clarification of property rights over environmental services.	Increased competition for control over land, social tensions.	Poorer farmers have equal rights in Small Groups. Small Group leadership supposed to change on a weekly basis.
Increased organizational and management capacity of community-based organizations through collaboration in delivery of environmental service.	Splitting up of communities in households gaining and households loosing due to the market. Potential threat to local culture as environmental services become commercial assets.	Participation in TIST Small Groups is voluntary and open for everybody. TIST Small Groups develop "best practices" (e.g., for starting nurseries and caring for the seedlings, for germinating and transplanting different tree species). Other TIST groups are encouraged to adopt them. Carbon sequestration put in wider context.
	Human Assets	
Improved education and skill base relating to forest management for environmental services, cooperation project management, marketing, negotiation, enterprise development, etc.	Poorer households excluded, can only work in subordinate jobs, necessary skills are brought in from outside.	Best practice development as education effort of the program. Improvement of local skills, training of 'Quantifiers' traveling to various locations within the program and record group results.
Improved health due to more varied diets, improved water supply (both quantity and quality), improved air quality and microclimate, increased disposable income for medical treatment.	Reduced health due to lost access to NTFPs and associated nutrition and reduced disposable income.	TIST addresses health issues (HIV/AIDS, malaria), education, and nutrition during Small Group meetings. Groups develop orphan care strategies.
	Physical Assets	
Investment in improved communication links to remote forest areas.	Improved communication channels only available to certain market participants leading to few spin-offs for the rest of the community, inequality increases.	TIST creates a communication and administrative structure. Palm computers and Global Position System (GPS) technology used by farmers to collect data on tree growth (and carbon storage). Data transmitted by internet.
	Financial Assets	
Diversified income base, regular income with prospect to be paid over many years.	If markets regulations exclude poorer farmers this leads to negative repercussions on their financial assets.	Cash payments for every living tree are rewarded regularly on bank accounts to be opened by Small Groups for this purpose.
	Political Assets	
Improved community-based organization provides basis for gaining political representation and voice.	If markets lead to marginalization of poorer farmers, these have even fewer channels to influence political decisions.	TIST encourages farmers to contribute to the continuous program improvement by submitting monthly reports and best practice proposals. Local democratic decision making of small groups is encouraged.

6.2. Potential Impacts on Other Groups

In chapter 2.4 we assessed several prospects as well as difficulties resulting out of the aim to link environment and development agendas in PES schemes and the CDM (i.e., the aforementioned twin-objective of the CDM). In the discussion of the main aspects to consider, also possible impacts on *other* groups, not directly participating in the project activity, were addressed. Other social groups besides the actual 'service providers', i.e., the farming households actually planting trees for the purpose of carbon sequestration can be affected by emerging environmental service markets. We differentiated between three main groups. Namely, poor service users, e.g., in the case of watershed protection where downstream peasants depend on the service provision of their upstream colleagues, secondly, poor landless laborers not engaged in farming activities prevailing in the project area and, thirdly, poor consumers.

With regard to the first group, a number of positive spillover effects of the project are identified. First and foremost the general finding of GRIEG-GRAN et al. (2005) is true for the TIST program as well. GRIEG-GRAN et al. (2005) highlight that the climate mitigation activity of project participants (i.e., the tree planting activity in the case of TIST) in fact facilitates adaptation to climate change of other poor farm households. There is no denying that non-TIST households profit from the positive, climate change mitigating impact of carbon sequestration on a macro scale. On the community scale, increased precipitation, wind breaks, shade provision, erosion control etc. positively affect the micro climate of the village and thus also benefit the 'external' service users (see e.g., LOS et al. (2006)).

The second identified group to look at, poor landless laborers, actually gain from the TIST project as well because it increases the general employment opportunities through the labor-intensive tree planting activity. The field survey revealed two interesting aspects in this regard. On the one hand, farmers participating in the project hired external labor, i.e., created employment opportunities for landless laborers. On the other hand, the Small Group structure of the TIST program enabled farmers to participate in the activity that could not spare any of their own land resources to participate in the program. Thus, the participatory Small Group structure actually enabled some of the poorer farm households to overcome their land constraint.

For poor consumers, the third group of interest, GRIEG-GRAN et al. (2005) show that setting aside a relatively large area for the program activity drives up prices for staple food regionally because of competition for land. If the emerging market for an environmental service provision drives up staple food prices, this would adversely affect poor urban consumers of this crop not able to produce the crop on their own property. For PES programs in general PAGIOLA et al. (2005) however point out that such effects appear quite unlikely. The total size of the areas enrolled in PES programs is simply too small to affect prices on a regional or even national scale. We found the same to hold true for the A/R carbon finance projects actually registered by the Executive Board of the UNFCCC (see chapter 2.4.). Depending on the actual scale of the PES or CDM LULUCF project activity, however, the described effect on food prices is an important point to look at. In the case of TIST, neither the farm household interviews nor the focus group discussions with village leaders on the community level did uncover such rather undesired effects of the program activity on staple food prices.

Thus, the overall impact of TIST's tree planting activity on 'external agents' not directly involved in program implementation is positive under this rather qualitative view with regard to rural development.

7. Conclusions and Policy Recommendations

The CDM framework as designed by the international community is unique and unprecedented. It is a scheme aiming to ensure development benefits while simultaneously creating an *additional* reduction of GHGs from the atmosphere. This twin objective to protect the global climate *and* to contribute to sustainable development is as challenging as it is simultaneously appealing. Especially the land-use related CDM and, thus, project activities in the forestry sector have a substantial potential for the development of rural areas. LULUCF projects that enhance the sequestration of carbon in biomass directly deal with the management of the resources on which the livelihoods of many poor people depend. The most fragile and threatened natural resources in developing countries and serious rural poverty share very often overlapping areas (SUNDERLIN et al. (2005)). Thus, apparent synergies between rural development, poverty reduction, and carbon sequestration in land-use related activities suggest that the aforementioned twin objective of the CDM can be reached if activities are designed accordingly.

However, there is clear need for a more concise conception of which development goals LULUCF projects aim to accomplish. Development benefits can occur on several different levels, e.g., positive effects for project participants, positive effects for both, participants and non-participants, or even wider development benefits for the host country. Without a clear conception of those different levels and the envisaged benefits, the synergistic goal of poverty reduction and resource conservation through the emerging market mechanisms of the CDM remains vague and sometimes just ambitious (TSCHAKERT (2007)). BARRETT et al. (2005) caution that '*synergies do not naturally emerge just because rural poverty reduction and renewable natural resource conservation are each appealing goals with common drivers and some intrinsic interlinkages.*' A particularly thoughtful design of the institutional structures is essential to foster rural livelihoods *and* natural resource management.

The results of the field research undertaken for this study emphasized this aspect in a very crucial point, namely the design of the actual carbon finance payment mechanism. As outlined above, the program aimed to channel the carbon finance payments to the participating local villagers through a voucher-based system by using a network of rural banks throughout Tanzania. However, bank accounts to deposit the program's vouchers could often only be opened in larger rural villages, while farmers exhibited high immobility due to prohibitive opportunity costs of transport in remote rural areas. As a result, the huge majority of farmers very seldom left the "home village" and therefore criticized that vouchers do expire and can no longer be deposited. Nine out of ten farmers were found to visit the next larger village with a rural bank branch only once per year. This was most often during Christmas time with the purpose to visit relatives. However, none of the banks operating in the research area was open during Christmas. Therefore, farmers were often unable to receive the payment they were rightfully expecting for their activities. The result was frustration and an increasing lack of trust in the program design and its institutions. Thus, the actual trickle-down of the carbon-related payments, the payment channels, must be very carefully designed to address the immobility of the farm households in remote rural areas. Clearly, the persons in charge for program design should be keen to assess mobility-related questions when setting up a carbon finance operation and its payment scheme. This might even include a random household survey to find out mobility patterns of the rural households and related costs of transportation etc. in order to set up an adequate payment system. In addition, non-expiring vouchers would allow for more time and flexibility for the household members to deposit the voucher at the nearest local bank at their convenience. Besides the aspect of fairness, it is clearly in the interest of the project's longevity to remunerate participants for their actual efforts. And, as long as the voucher is not deposited the accruing interest is a capital gain of the project proponent. Thus, there is economically no harm in issuing non-expiring vouchers to program participants. Furthermore, the payment channels, i.e., the network of local banking institutions have to be scrutinized. Possibly, local 'Savings and Credit Cooperatives' (SACCOS), the most common micro-financial organizations in rural Tanzania, that are present even in smaller villages could be used to disburse carbon finance payments. As described in chapter 3.3.3, SACCOS institutions are often located near their members' workplaces or homes. This effectively reduces transaction costs for farmers and makes SACCOS' services in turn much more attractive than other government or private banks.

However, the other side of the coin is that SACCOS' employees not often have an appropriate banking background so that financial and general management decisions have been reported to be negatively affected (see chapter 3.3.3). The bottom line is that project proponents might not even be able to rely on existing financial structures to disburse carbon finance payments but might have to develop and establish a project-specific infrastructure. However, such effort can become logistically complicated, challenging from a security perspective and very time-consuming on top. In addition, the design of an appropriate payment system includes ascertaining that project benefits are distributed within the Small Groups. The formation of such groups to counteract high transaction costs is a very appropriate strategy. However, clear competencies and rules of leadership within the group should be developed and put down in order to ensure that payments also trickle down in the group. Even within the group, underprivileged farmers are often not in a position to express their interests. This clearly bears the risk that better endowed farmers become more assertive, dominate the group's interests, and ultimately usurp payments. Thus, without a doubt it is one of the most challenging efforts in the design of a carbon finance or PES program to implement a functioning payment system, which ultimately guarantees the program's success.

Another issue, identified regarding program implementation, was the rapid expansion during the program's pilot phase. The geographical scatteredness and the remoteness of participating farm households made it almost impossible to adequately track and take care of the ongoing planting activities. Particularly the starting phase of such a new program has its ambiguities for project proponents. A quick uptake of the new activity creates the momentum needed to ensure a critical starting level of participation. But a rather steep growth of participation leads inevitably to rather unfavorable 'bushfire phenomena'. Chapter 3.1 mentioned exemplarily such a situation reported in the Chinese 'Grain for Green' program (XU et al. (2004)). Almost unavoidably, after an initial surge in participation, a subsequent shrinking process leads to disappointment and frustration on both sides, the implementing farmers as well as program proponents. Moreover, in the case of TIST participants that dropped out during the first program phase challenged the fundamental program idea and showed reluctance to engage in similar initiatives in the future when asked in the field survey.

However, besides the challenging issues regarding payment design and program expansion, the applied structural equation modeling approach was able to assess a number of remarkable factors determining participation in the program and to identify their impact on the livelihood situation of the farm households. With regard to the poverty-relevant aspect of the tree planting activity, a very important finding was that the program design in fact promoted farmers with a rather distinct land constraint to actively participate. The major reason for that lies in the Small Group structure of the program. The Small Groups enabled farmers to plant trees collectively, for example on a larger farm of one of the group members. The carbon finance revenues generated through that activity are then supposed to be shared within the entire Small Group. The importance of this finding becomes evident by looking at other PES program structures in comparison. We saw that the emerging CDM market can become inaccessible for poorer farmers if land tenure rights are uncertain or even lacking completely. GRIEG-GRAN et al. (2005) reported a large proportion of farmers to be often excluded from environmental payment schemes due to exclusive entry regulations like minimum land size or explicit land title requirements. In addition, subsistence needs of the farm household conflicting with the tree planting activity can lead to a biased program participation of better-endowed households. However, findings of the former chapters reveal resource-poorer farmers to exhibit a greater inclination to participate in the TIST program than their better-endowed colleagues as approved in the SEM analysis. Resource endowments, market incentives, and bio-physical farm characteristics showed a rather weak effect on the household's inclination to participate in the carbon finance program.

Thus, in contrary to the study of PATTANAYAK et al. (2003) who found those assets to be a significant driver of adoption in most of the studies they reviewed, we saw that those factors were only of minor importance for the participation decision in TIST. The SEM analysis shows that the program does not discriminate against people who have less land than others. A very encouraging finding, with regard to the CDM's twin-objective as a possible land constraint can be relaxed, which in turn supposedly leads to higher productivity and income. The same is in fact true for possibly existing labor constraints within the household. With regard to labor availability for agricultural production, we saw that almost 10% of all interviewed households had to either hire external labor capacities or count on labor availability in the 'extended family' to counteract seasonal labor shortages. Labor capacity in a subsistence environment is of particular interest to assure food production for the farm household in times of scarcity. Here again, the Small Group structure of the program can become a safety net for the individual farm household because labor resources can be shared uncomplicatedly amongst farmers of the group, which counters possible labor constraints.

Furthermore, we saw that the program's planting activities (e.g., through planting of windbreaks and hedgerows along property lines) promote the formalization and clarification of resource tenure rights. This is particularly advantageous as tenure security promotes a long-term carbon storage engagement. Field research revealed that farm households held only for approx. 3% of all plots they cultivated an official certificate of ownership. Even though a lack of titling should not be equated with complete tenure insecurity given the fact that customary regulations often prevail, the TIST program encourages farmers to plant trees around homes, farms, roads, and alongside paths to set clearer signs of existing borders. The value added of this activity cannot be overestimated as it clearly contributes to more tenure security with regard to the resource base on which the farm household depends.

Above, we outlined one major positive aspect of the joint afforestation efforts by the Small Groups, namely that the program does not discriminate against people who have less land than others do. However, a word of caution shall be added as lacking responsibility and access control over areas where TIST trees have been collectively planted was found to be a major problem. For the carbon storage provision, it is crucial to control livestock grazing and hence browsing damage to the trees planted by the project. Yet, given the fact that livestock keeping is a major source of income, especially in the lowland areas of the research area and, moreover, taking into account that livestock is a particular livelihood reserve in times of hardship, the project should actively promote the introduction of so-called 'silvopastoral' systems. The rationale for this recommendation is that our household survey found TIST farmers to be even better endowed regarding livestock than non-TIST farmers (see chapter 3.3.2).

Silvopastoral systems in essence integrate tree management and livestock keeping. They bring together different tree and shrub species with fodder plants such as legumes and grasses for animal nutrition. PAGIOLA et al. (2007) distinguish between three different forms of silvopastoral systems, namely, planting trees and shrubs in pastures, or using them for fencing, controlled grazing of livestock within tree plantations and so called 'cut and carry systems', in which livestock is fed with the foliage of planted trees and shrubs ('fodder banks'). In the latter system, fodder supply to feed the livestock can also be taken out of the system (i.e., spatial disintegration). All the systems have in common that they combine multiple purposes including soil amelioration through nitrogen fixing tree species, provision of shade, fodder, fruit, wood production and carbon storage (DAGANG and NAIR (2003)).

Most importantly, thoroughly planned silvopastoral systems are not only able to reconcile the tree planting and the livestock keeping activity of the farm household but additionally improve the farm household's cash flow situation. In essence, two short-term but periodically reoccurring income sources from forage and livestock as well as carbon certificate sales are combined with the long-term investment of the tree-planting component. This multi-product character of the silvopastoral system 'insures' the farm household against unfavorable market situations of either one of the products, adverse weather conditions, or agricultural policy decisions (SHARROW (1999)). In addition, research results show that silvopasture can be realized on small farms as well as on large landholdings, as either 'stand-alone operation', or as part of different land-uses that comprise pasture systems and diverse timberlands, thus making it most suitable for the diverse land-use situations found in the research area (NOWAK et al. (2007)).

Yet, another important aspect of the field research and the subsequent structural equation modeling should be emphasized here. We found that social interactions and the connectedness of the farm household within the village were of major importance regarding the inclination to participate in the carbon finance program and the tree planting activity. NARAYAN (1997) found those interactions, described as *'horizontal ties in associational life'*, or social capital, to positively impact on household welfare. The concept of social capital introduced by PUTNAM (1995) was concretized in the area of development research as networks, that is social relations with a particular set of rules, norms, obligations, and reciprocity that determine collective action and decisions. Social capital therefore plays a crucial role in enabling the network members to achieve their individual and community development goals (WOOLCOCK (1998); NARAYAN and PRITCHETT (1999); NARAYAN (1997)). Our model in chapter 5 clearly revealed that the latent variable depicting social interactions of the farm households had the highest impact on the household's inclination to participate in the carbon finance program. This is insofar of particular relevance as it confirms that economic models based merely on profit or utility maximization fall short of taking attitudinal and social variables, determinants of the decision process, adequately into account. The aforementioned was originally found by ROGERS and SHOEMAKER (1971), and is according to our modeling of particular relevance in a developing country's context. Variables depicting social capital are fundamental to explain household decision making processes with regard to adoption.

Interestingly enough, PATTANAYAK et al. (2003) in their meta-analysis of agricultural adoption studies found that variables measuring social capital were hardly included in any of the studies they reviewed. In this regard, it is important to note that in order to achieve satisfying explanatory power, social capital related aspects must be emphasized in future work besides the more common asset-related variables, variables measuring credit availability, or price proxies. That results from the fact that social interactions enable the farmers to discuss both, positive as well as negative aspects of the carbon finance program regarding its impact on their livelihoods. Social interactions have a risk-mitigating effect since better-connected households are more capable to adapt to perceived risks of the new technology because of further information sources provided by their networks. Therefore, better-connected households, that is, households with more social capital, have a greater inclination to participate in the program. A direct policy implication is the further strengthening of social networks through the Small Group aspect of the TIST program. In this regard, NARAYAN (1997) cautions that solely creating community groups generates neither trust nor social capital. However, the appealing aspect of the TIST program is that the Small Groups were formed in a process that encouraged rural farmers to identify their primary needs and subsequently to develop ideas how to overcome the major shortcomings in the livelihoods of their families. As described in chapter 3, the process resulted in 40 mutually responsible Small Groups of farmers to prevent the risk of famine and to reduce poverty. Additionally, farmers were convinced of the joint effort to reforest degraded lands to improve their livelihood situation. NARAYAN (1997) recommends that community-determined needs should lead to the formation of community-based organizations. We found that this situation fully applies in the case of the TIST program, possibly explaining its success in parts and thus strongly suggest that self-determined processes, strengthening social networks, and the formation of community-based organizations should be actively promoted in PES or CDM operations. They provide the fundamental basis for gaining political representation and voice of the farmers and ensure the longevity of the program activities and goals.

A general concern in development studies is that social groups might become exclusive, be it from a gender perspective or with regard to different religious affiliations or denominations. NARAYAN (1997) reaffirms this concern in a Tanzanian context. In her study on social capital, she found considerable gender differences in community groups and thus cautions that development benefits must equally reach women as they must reach men. However, in the context of TIST the structural equation model showed that more male household members had a rather negative influence on the participation decision of the household. Our initial reasoning above was that a higher risk-tolerance of male household members and the greater labor availability of male-headed house-holds would positively influence the adoption decision. However, the contrary is true in case explored during field research. We observed significantly more female-headed households within the group of TIST farmers compared to their non-TIST colleagues. Clearly, female-headed households show a higher self-motivation to participate in the program. The long-term aspect of the carbon finance revenues is attractive for female-headed households given their challenging task to reconcile family and childcare with farming activities. This finding is of particular interest because we also hypothesized earlier that female-headed households trying to balance various different activities do not have the possibility to try out a new technology.

The tree planting activity with the particular carbon finance incentive and its long-term perspective, however, stabilizes the livelihood situation of female-headed households. Thus, from a development perspective, it is helpful to encourage the participation of female-headed households. At the same time, the program should actively encourage the participation of male-headed households in order to ensure the programs gender balance. The same aspect of balance is true for religious affiliations. We saw that Anglican denomination of the household head positively influenced the household's inclination to participate in the program. Given the fact that the Anglican Church initially promoted and fostered the program's starting phase this was not unexpected. However, one of the main objectives of the TIST program is to broaden its outreach. Therefore, all interested farm households, regardless of denomination or religious affiliation, should have a chance to participate. Given the study results, it is worth to look at the future targeting of the program in order not to narrow program participation a priori by only addressing specific groups of the society, e.g., a specific denomination.

With regard to the periodically recurring income effect sought by farm households participating in the afforestation activities we saw that more than two-thirds of farm households would be willing to commit themselves to a rather long-term participation of up to several decades. Field research findings show that higher uncertainties in the farm environment positively influence the household's inclination to participate in carbon finance-related activities. Therefore, we conclude that farm households perceive participation in the program as a risk mitigation option. The reoccurring carbon finance payments provide a continuous stream of revenue, which helps to smooth farm income fluctuations. The specifically engaging characteristic of those sequestration service payments is that they help to reduce the long gestation period of a tree planting activity and provide returns even prior to maturity of the trees for other purposes. Furthermore, it should be considered to adapt the actual contract length to the often rather short-term needs and planning horizon of the farm households. Our findings show that it would be most appropriate to offer a stable, long-term program framework but to think of flexible contractual solutions within this framework. A single-sided flexibility giving the farm household the opportunity to renew a five-yearly contract periodically was most attractive in this respect.

In summary, we saw that tree planting, as a land use-related climate mitigation activity, had a very broad and positive impact on the asset base on which subsistence farmers depend. This includes not only natural assets, but also social and human assets. Our initial hypothesis with regard to the importance of social interactions has proven to be true. Social capital is of utmost importance for the participation decision in land use related project categories of the CDM. With regard to the CDM's twin-objective, seen from a broader perspective, it is important to note that the current CDM focus on only afforestation and reforestation project activities in developing countries as a means for climate mitigation and sustainable development is clearly too narrow. Current UNFCCC regulations only appreciate services related to carbon sequestration as active storage of carbon in newly established trees. While carbon conservation, as for example the protection of an existing carbon reservoir like native forests is not rewarded as a climate-relevant service provision. It is a clear shortcoming of the current international climate regime that neither the UNFCCC nor the Kyoto Protocol include a satisfying mechanism to reduce the substantial climate-relevant emissions from deforestation and forest degradation, which are responsible for about a quarter of the global greenhouse gas emissions (DEFRIES et al. (2002); HOUGHTON (2003)). Just recently government-backed initiatives addressing 'reduced emissions from deforestation and degradation' (REDD) in the context of the international climate regime begin to emerge (STERN (2006); UNFCCC (2006a); THE WORLD BANK (2007)). It is a politically challenging task for the next years to come to fully integrate the protection of carbon stocks of forest ecosystems into the existing climate regime.

7.1. Outlook – Limitations and Further Research

The analysis presented has shed more light on the interlinkages between climate mitigation activities in the area of land use, land-use change, and forestry (LULUCF) and the potential for development of rural areas. The rural household was consequently at the center of the analysis with the major aim to assess factors that determine the participation of smallholders in LULUCF carbon finance projects. Chapter 7 has put forward policy-oriented advice with regard to the design and development of LULUCF projects under the CDM in the future. We saw that thoughtfully designed land use-related CDM projects are able to combine both, aspects of climate change mitigation and rural development as they have a positive impact on a very broad asset base on which poor farmers depend. This included, but was not limited to natural assets, human assets, and social assets. At the outset we hypothesized that social interactions expressed by the connectedness of the farm household within its village environment, would be of utmost importance for the participation decision in land use related project categories of the CDM. This hypothesis was confirmed by structural equation modeling. This methodology enabled us to assess factors that determine the participation of smallholders in LULUCF carbon finance projects in more detail.

Structural equation modeling (SEM) offered a suitable approach to detect and investigate underlying factor categories that influence the participation decision of smallholders. The advantage of the methodology applied was that it is not limited to model only measurable or directly observable variables. In contrary, SEM is ideal to reveal underlying root causes of an adoption decision by also modeling latent variables that determine observable and therefore measurable outcomes. In this regard, we identified a major shortcoming of common agricultural adoption studies, namely a rather biased focus on issues related to farm resource endowment, which ultimately disregards aspects of social capital. Social networks play a major role for adoption decisions in rural areas. Particularly in rural Tanzania, these social networks control adoption decisions what might be overlooked if the respective variables are not systematically surveyed and taken into account in future studies. Still, many aspects remain unknown regarding social capital in different social and economic contexts. However, the factor must to be included in the household welfare equation in Tanzania. In this regard, it would be interesting from a scientific point of view to further explore new and different indicators, i.e., manifest variables that might be proxying for different aspects of social capital during future household surveys. This would not only help to further investigate the aspect of social capital from innovative and different angles but would be beneficial for the other factor categories as well. While we constructed our factors in chapter 5 based on variables already discussed in the literature, a way forward would be to think about and explore new proxies relying on the fact that the underlying latent constructs turned out to be valid. Thus, the arena for further research would be determined by the latent constructs while their actual proxies should be explored more innovatively leading research into new aspects of adoption.

In addition, a rather long-term outlook of the socio-economic impacts land use-related CDM projects have for subsistence farmers requires on the one hand longitudinal data and on the other further research in the realm of impact evaluation. The first limitation reflects the 'snapshot' aspect provided by the cross-sectional survey data underlying this study. We saw that with regard to prices, for example, cross-sectional data had various constraints. Per definition, it is time invariant and thus unable to reveal market fluctuations. Furthermore, household survey data is usually self-reported and thus often exhibits a lack of statistical variability. The fact that agricultural household surveys represent only a snapshot in time often goes back to time and budgetary constraints in the research design. We stuck to conservativeness with regard to our field research data and instead of deliberately extending it via longer recall periods in the survey to a longitudinal setting, we strongly recommend additional and future research in the same or similar settings allowing to ultimately base research findings on time series data.

With regard to a more comprehensive impact evaluation approach BAKER (2000) suggests to base research on four different parts. Namely, comprehensive monitoring, process evaluation, cost-benefit evaluation, and impact evaluation. While project monitoring merely assesses whether a program is being implemented as it was designed, process evaluation is of more scientific interest. The latter technique evaluates how the program operates and focuses on problems in environmental service delivery. The third aspect of impact evaluation is cost-benefit or cost-effectiveness evaluation. Program costs (i.e., monetary and/or non-monetary ones) are assessed under particular consideration of alternative uses of the same resources and compared to the benefits being produced by the program. Finally, of most scientific interest for future research is the aspect of impact evaluation in a stricter sense. In general, impact evaluation intends to determine whether the program had the expected and planned effects on individuals or households. Its goal is to thoroughly examine whether those effects are directly attributable to the program intervention. In doing so, impact evaluation can also unveil unintended consequences of the program implementation, whether positive or negative. The interesting feature of strict impact evaluation is that it attempts to net out the project's outcome from other factors or events that are correlated with those outcomes but are not caused by the project itself. The procedure tries to answer the question what would have happened had the project never taken place. Scientifically the so called 'counterfactual' has to be estimated (BAKER (2000)).

Of particular interest for further research is the aspect how the rise of markets for environmental services impacts on the poor on a longer term and what impact these markets have on their welfare development. Exploring this question in more detail is not only of concern for ethical reasons but of particular scientific and political interest because marginalization of the poor from the market development process would have critical implications for the sustainability of market systems, and ultimately undermine their effectiveness in achieving environmental objectives.

App. 1 Residual Covariances for Initial Model

	tist	rest	resi	flbr	soil	slope	fasi	vsts	memb	n_ext	infc	prce	madi	incm	cred	a_li	s_exp	a_hh	dsct	educ	age	gndr
tist	0																					
rest	0	0																				
resi	0,044	-0,178	0																			
flbr	0,007	-0,054	0,124	-0,001																		
soil	0,001	0,026	0,083	0,07	-0,022																	
slope	0,002	0,035	0,1	0,097	-0,008	0																
fasi	-0,003	-0,048	0,085	0,077	-0,008	-0,001	0															
vsts	0,017	0,043	-0,102	-0,018	-0,076	-0,095	-0,061	0														
memb	0	0	0,053	0,053	0,003	-0,001	0,004	0	0													
n_ext	0,037	-0,041	0,037	0,076	0,088	0,069	0,11	-0,1	0,009	0,013												
infc	0,008	-0,05	0,08	0,057	0,095	0,096	0,077	-0,052	0,001	0,012	0,001											
prce	-0,001	-0,003	-0,06	-0,001	0,01	-0,007	-0,002	0,034	-0,006	-0,003	-0,044	0										
madi	-0,007	0,468	-0,137	0,064	0,015	-0,042	-0,147	0,068	0,212	-0,015	-0,202	0,01	0									
incm	-0,023	1,048	-0,049	0,106	-0,023	-0,054	-0,05	0,1	0,605	0,993	0,887	0,017	0,498	0								
cred	-0,006	-0,017	0,071	0,008	-0,041	-0,033	0,001	-0,109	-0,032	0,038	0,016	0	0,001	-0,069	0							
a_li	0,091	-0,511	0,194	0,289	0,056	0,338	-0,048	-0,171	-0,121	0,24	0,299	0,662	1,341	-2,762	-0,826	0						
s_exp	-0,009	-0,008	0,054	0,083	0,021	0,018	0,011	-0,13	0,005	0,15	0,107	-0,006	0,039	0,995	0,031	-0,084	0					
a_hh	0,01	0,041	0,056	0,099	-0,077	-0,014	-0,057	-0,237	-0,002	0,245	0,175	-0,011	0,272	1,272	0,021	0,471	-0,01	0				
dsct	0,012	-0,157	0,169	-0,002	0,133	0,172	0,152	-0,037	0,112	0,169	0,093	-0,02	0,307	0,394	0,049	0,793	0,119	0,205	-0,004			
educ	0	0	-0,022	0,001	0,001	0,002	-0,002	-0,016	0	0,014	0,006	0	0,025	0,152	0	-0,106	-0,003	0,012	-0,008	0		
age	0,006	-0,005	0,007	0,047	0,029	0,048	0,026	-0,015	0,021	0,01	0,022	-0,023	0,08	-0,006	0,034	0,213	0,022	0,048	0,103	-0,002	0,001	
gndr	-0,029	0,037	-0,035	-0,01	-0,017	-0,016	-0,02	-0,008	-0,018	-0,024	-0,013	-0,007	0,052	0,305	0,022	-0,08	-0,008	0,005	0,006	0	-0,002	0

App. 2 Standardized Residual Covariances for Initial Model

	tist	rest	resi	flbr	soil	slope	fasi	vsts	memb	n_ext	infc	prce	madi	incm	cred	a_li	s_exp	a_hh	dsct	educ	age	gndr
tist	0																					
rest	0	0																				
resi	1,382	-1,728	0																			
flbr	0,421	-1,051	4,25	-0,036																		
soil	0,035	0,239	1,371	2,31	-0,247																	
slope	0,066	0,304	1,567	3,04	-0,098	0																
fasi	-0,111	-0,499	1,609	2,885	-0,113	-0,012	0															
vsts	0,47	0,378	-1,601	-0,576	-1,138	-1,35	-1,039	0														
memb	0	0	0,835	1,664	0,042	-0,013	0,073	0	0													
n_ext	1,055	-0,366	0,577	2,382	1,328	0,98	1,896	-1,384	0,123	0,136												
infc	0,434	-0,809	2,321	3,274	2,626	2,53	2,443	-1,344	0,031	0,297	0,037											
prce	-0,058	-0,035	-1,299	-0,025	0,205	-0,136	-0,037	0,654	-0,121	-0,064	-1,595	0										
madi	-0,088	1,889	-0,985	0,916	0,102	-0,278	-1,155	0,441	1,379	-0,099	-2,443	0,087	0									
incm	-0,106	1,499	-0,125	0,541	-0,055	-0,126	-0,14	0,229	1,39	2,307	3,79	0,052	0,526	0								
cred	-0,176	-0,153	1,113	0,263	-0,621	-0,462	0,01	-1,523	-0,441	0,536	0,415	-0,005	0,006	-0,153	0							
a_li	0,547	-0,954	0,646	1,918	0,178	1,015	-0,174	-0,515	-0,361	0,728	1,669	2,748	1,858	-1,354	-2,474	0						
s_exp	-0,28	-0,079	0,989	3,004	0,356	0,283	0,201	-2,107	0,081	2,469	3,255	-0,133	0,292	2,654	0,502	-0,285	0					
a_hh	0,245	0,307	0,756	2,654	-0,959	-0,168	-0,805	-2,848	-0,024	2,981	3,926	-0,179	1,517	2,508	0,249	1,183	-0,123	0				
dsct	0,33	-1,379	2,633	-0,051	1,998	2,441	2,593	-0,519	1,579	2,413	2,434	-0,397	1,996	0,908	0,696	2,386	1,952	2,491	-0,039			
educ	0	0	-1,312	0,177	0,081	0,095	-0,134	-0,868	0	0,756	0,55	0,033	0,605	1,329	0,001	-1,205	-0,209	0,534	-0,427	0		
age	0,541	-0,143	0,335	4,858	1,442	2,236	1,451	-0,68	0,968	0,464	1,869	-1,481	1,717	-0,048	1,597	2,105	1,194	1,911	4,801	-0,305	0,116	
gndr	-2,354	0,949	-1,592	-0,889	-0,736	-0,67	-1,006	-0,325	-0,757	-0,996	-1,026	-0,405	0,983	2,057	0,907	-0,701	-0,403	0,164	0,238	0	-0,258	0

App. 3 Initial Model: Modification Indices Regression Weights

			M.I.	Par Change
tist	<---	gndr	4,416	-0,207
resi	<---	flbr	6,963	0,349
flbr	<---	resi	12,86	0,072
flbr	<---	infc	4,435	0,072
flbr	<---	s_exp	4,907	0,047
vsts	<---	rend	11,003	-0,128
vsts	<---	resi	7,66	-0,167
vsts	<---	flbr	4,508	-0,255
vsts	<---	s_exp	8,125	-0,181
vsts	<---	a_hh	13,909	-0,175
memb	<---	madi	4,207	0,046
infc	<---	biof	6,328	0,102
infc	<---	rend	12,049	0,082
infc	<---	flbr	6,289	0,186
infc	<---	soil	5,991	0,087
infc	<---	slope	6,263	0,084
infc	<---	fasi	4,309	0,084
infc	<---	madi	6,206	-0,039
infc	<---	incm	10,819	0,018
infc	<---	s_exp	8,565	0,114
infc	<---	a_hh	11,079	0,096
madi	<---	infc	4,684	-0,602
madi	<---	dsct	4,728	0,326
madi	<---	age	4,493	1,046
incm	<---	rend	7,502	0,789
incm	<---	n_ext	5,638	0,974
incm	<---	infc	12,318	2,648
incm	<---	s_exp	7,664	1,311
incm	<---	a_hh	6,869	0,917
incm	<---	gndr	4,032	2,388
cred	<---	infc	4,489	-0,147
a_li	<---	minc	9,003	1,961
a_li	<---	prce	8,748	1,304
a_li	<---	incm	5,217	-0,118
a_li	<---	cred	7,255	-0,857
a_hh	<---	vsts	4,361	-0,133
a_hh	<---	infc	5,587	0,28
a_hh	<---	a_li	4,586	0,029
dsct	<---	madi	5,186	0,043
age	<---	flbr	13,985	0,156
age	<---	madi	6,48	0,022
age	<---	dsct	16,222	0,076
gndr	<---	inpa	5,828	-0,114
gndr	<---	tist	5,828	-0,114

App. 4 Household Questionnaire

The information provided herein will be kept confidential and will only be used for academic purposes

Project on Carbon Finance and Rural Development in Tanzania

Agricultural Household Survey

Center for Development Research, University of Bonn, Germany
Institute for Agricultural Policy and Market Research, Justus-Liebig-University, Giessen, Germany
TIST – The International Small Group and Tree Planting Project, Morogoro, Tanzania
Sokoine University of Agriculture, Morogoro, Tanzania

ALL FIELDS IN LIGHT GREY TO BE FILLED OUT BY ENUMERATOR

Part 0. General Information

Questionnaire Number					
Enumerator Name					
Date of Interview					
Time taken		Start		End	

Data Checked by	Date	Status		Comments
		OK	Return	

Data Entry	Name	Date	Comments
Pages			
Pages			

Household Identification		
Region		
District		
Ward		
Village		
Household No.		

Head of Household
Address
Name of the Head of the Household

INTERVIEWER: PLEASE READ THE FOLLOWING INSTRUCTIONS CAREFULLY!

INTRODUCE YOURSELF AND ASK FOR A PERSON THAT CAN ANSWER THE QUESTIONNAIRE [**HOUSEHOLD HEAD**].

IF THE PERSON IS NOT AVAILABLE, ASK WHERE YOU CAN FIND HIM/HER OR IF YOU COULD COME BACK LATER. IF YOU CAN LOCATE HIM/HER INTRODUCE YOURSELF. EXPLAIN THE PURPOSE, OBJECTIVES, AND EXPLAIN THE PROCEDURE. ASK WHETHER THE FARMER WOULD BE WILLING TO PARTICIPATE IN THE RESEARCH.

IF THE RESPONDENT IS AVAILABLE NOW ASK THE FOLLOWING QUESTIONS:

Are you owning a piece of land where you do agriculture?

☐ **1 – YES**
0 – NO

What is the size of your lot/piece of land/farm? **1-ACRES** **2-HECTARS** **0 -OTHER (__________)**		[IN CASE HE/SHE DOESN'T OWN LAND] What is the size of the lot you are working at? **1-ACRES** **2-HECTARS** **0 -OTHER (__________)**	
Quantity	**Code**	**Quantity**	**Code**

How many families work at this lot/piece of land/farm?

[E.G. THE HOUSEHOLD HEAD PLUS THE WIFE AND SINGLE SONS/DAUGHTERS WOULD BE FAMILY 1, THE SON AND/OR DAUGHTER OF THE HOUSEHOLD HEAD LIVING ON THE SAME LOT/PIECE OF LAND WITH THEIR FAMILY WOULD BE FAMILY 2]?

☐ **QUANTITY**

Do the families that live in this lot always work together and share the products?

☐ **1 – YES**
0 – NO

IF <u>YES</u>, USE <u>ONE QUESTIONNAIRE FOR THE WHOLE FAMILY</u> [INCLUDING THE FAMILIES OF THE SONS AND/OR DAUGHTERS LIVING IN THE SAME LOT AS THE FATHER].

IF <u>NO</u>, CONSIDER <u>ONLY THE FAMILY OF THE RESPONDENT</u> OF THE VISITED LOT [EXCLUDING THE OTHER FAMILIES EVEN IF THEY LIVE IN THE SAME LOT].

Respondent	**Name**	**Surname**

☐

1 - HOUSEHOLD HEAD
2 – WIFE / HUSBAND OF HOUSEHOLD HEAD
3 – COHABITING PARTNER OF HOUSEHOLD HEAD
4 – SON / DAUGHTER OF HOUSEHOLD HEAD
5 – BROTHER/SISTER OF HOUSEHOLD HEAD
6 - PARENT OF HOUSEHOLD HEAD

ETHNIC GROUP OF THE HEAD OF THE HOUSEHOLD	RELIGION OF THE HEAD OF THE HOUSEHOLD	
	1 - MOSLEM	
	2 - CHRISTIAN	
	3 - TRADITIONAL	
	4 – OTHER (__________)	

Do you have any leadership role in the village / community?

☐ **1 – YES**
0 – NO

Which role do you have? ____________________

Part 1. Household Composition and Human Capital

A "household" includes all members of a residence that are sharing consumption of food and other items. Include workers or servants as members of the household only if resident at least six months in the household.
The next questions refer to all the household members, USE TABLE 1

TABLE 1

ID	1.What are the names of the people that live here with you?	2.The respective person is the [...] of the Household Head 1 -HOUSEHOLD HEAD HIMSELF 2 -WIFE / HUSBAND 3 - COHABITING PARTNER 4 -SON / DAUGHTER 5.- BROTHER/SISTER 6. -PARENT OR PARENTS-IN-LAW 7 -GRANDCHILD 8 -OTHER RELATIVES 0 -OTHERS (________)	3.Gender MALE FEMALE 1 0	4.Age	5.Marital Status 1 – SINGLE ADULT 2 – MARRIED 3 – SEPARATED / DIVORCED 4 – WIDOWED 5 – CHILD/PUPIL	6.Education 0 – NOT SCHOOL AGE 1 - NO FORMAL EDUCATION 2 – ADULT EDUCATION ONLY 3 - STANDARD I – VII PRIMARY SCHOOL 4 – FORM I – IV O - LEVEL 5 - FORM V – VI A - LEVEL 6 - BACHELOR 7 - MASTER/PHD 8 – (________)	7.Years of Schooling PLEASE TRANSFORM STATUS OF EDUCATION IN NUMBER OF YEARS OF SCHOOLING	8.What is the main occupation of ...? 0 - NONE 1 - FARMING 2 – SELF EMPLOYED (BUSINESS) 3 – PUBLIC SECTOR EMPLOYEE 4 – PRIVATE SECTOR EMPLOYEE 5 – FOREST RELATED EMPLOYMENT 6 - HOUSEWIFE 7 - STUDENT 8 – (________)
ID	Name	Code	Code	Years	Code	Code	Years	Code
1.								
2.								
3.								
4.								
5.								
6.								
7.								
8.								
9.								
10.								

11.								
12.								
13.								
14.								
15.								
	What are the names of the family members that don't live here with you?	ENUMERATOR PLEASE USE SAME CODING AS IN UPPER PART OF THE SAME TABLE						
		Relation to HH Head	Gender	Age	Marital Status	Education	Years of Schooling	Main Occupation
ID	Name	Code	Code	Years	Code	Code	Years	Code
16.								
17.								
18.								
19.								
20.								

Part 2. Land Use - ENUMERATOR: USE AN EXTRA SHEET
Please, let me ask you some questions about your lots/pieces of land. Please draw all your lots/pieces of land and show all the different lots and plants, e.g. annual crops and permanent crops [mixed crops and monocultures], pastures, woods, (yard, small garden), house, (small rivers), and so on. GIVE EACH LOT / CROP A NUMBER. MARK THE AREA ACTUALLY UNDER CULTIVATION AND THE DATE, WHEN THE CROP WAS PLANTED - USE **TABLE 2**.

PLOT CODE	1.What is the total area of your lots/pieces of land you are working at? 1 –ACRES 2 – HECTARES 3 – SQUARE METERS 0 -OTHER (______)		2.What is the primary use of that piece of land? 1 – AGRICULTURE 2 - HORTICULTURE 3 - TREE CROPS 4 – FOREST 5 – FALLOW 6 – PASTURE 7 – UNUSED 0 – OTHER (______)	3.What legal title or ownership rights do you have for this plot? 1 – LEASE / CERTIFICATE OF OWNERSHIP 2 – CUSTOMARY LAW 3 – BOUGHT 4 - RENTED 5 – BORROWED 6 – SHARE CROPPING 7 – OTHER (______)	4.For how long will you be able /allowed to use this plot? 1 – THIS YEAR ONLY 2 – THIS AND NEXT YEAR 3 – NEXT FIVE YEARS 4 – NEXT TEN YEARS 5 – NO RESTRICTION 6 – OTHER (______)	5.What crops have you grown on this plot during the last 12 months? SEE CROP CODES ON SEPARATE SHEET						6.How long does it take to get to your plot?
						SEASON ONE			SEASON TWO			
						MAIN CROP	2ND CROP	3RD CROP	MAIN CROP	2ND CROP	3RD CROP	
	Quantity	Code	Code	Code		Code	Code	Code	Code	Code	Code	Minutes
1.												
2												
3												
4												
5												
6												
7												
8.												
9.												
10.												

PLOT CODE	7.What soil type is this plot? 1 - SANDY 2 – LOAMY 3 - CLEYEY	8.In general, what is the slope of this plot? 1 - FLAT 2 - SLIGHT SLOPE 3 – MODERATE SLOPE 4 - STEEP SLOPE	9.If you were to sell this plot of land today, how much could you sell it for? ENUMERATOR PLEASE MARK ROW WITH A +++ IF PERSON CAN NOT EVEN THINK OF SELLING THIS LAND
	Code	Code	Tsh
1.			
2.			
3			
4			
5.			
6.			
7.			
8			
9.			
10.			

Part 3. Food – Consumption, Expenditure

1.What quantity of the crops mentioned above did you produce in the last agricultural year from January 2004 to December 2004?			2.How much of these crops did you sell last year?		3.Did your family also buy these products?	4.How much of these products did you buy last year?	
COPY CROP CODES FROM TABLE 2 ABOVE	1 – BAGS 2 - TINS		1 – BAGS 2 - TINS		1 - YES 0 - NO		
Code	Quantity	Code	Quantity	Code	Code	Quantity	Code

Part 4. Livestock

ANIMAL CODE	1.Does any member of your household currently own any of the following animals? FIRST ASK QUESTION 2 FOR ALL ANIMALS, THEN ASK QUESTIONS 3 FOR EACH ANIMAL.		2.How many [...] does your household currently own?	3.If you were to sell one of those [...] today, how much money could you get for it?
	Animal	1 - YES 0 - NO	Quantity	Tsh
1	Cows / Cattle			
2	Donkeys / Mules			
3	Pigs / Sows			
4	Sheep			
5	Goats			
6	Chicken / Ducks / Guinea Fowls			
7				
8				
9				

Part 5. Household Assets

EQUIPMENT CODE	1.Does your household own any of the following equipment? FIRST ASK QUESTION 1 FOR ALL ITEMS.THEN ASK QUESTIONS 2+3 FOR EACH ITEM BEFORE GOING TO THE NEXT ITEM.		2.How many [....] does your household own?	3.If you were to sell one of those [...] today, how much money could you get for it?
	Type of farm equipment	1 - YES 0 - NO	Number	Tsh
1	Car / Pickup / Truck			
2	Tractor			
3	Motorbike			
4	Bicycle			
5	Ox cart			
6	Small cart pulled by person			
7	Animal pulled plow			
8	Hoes			
9	Axes			
10	Rakes			
11	Shovels			
12	TV / Video			
13	Radio			
14	Bed			
15	Mattress			
16	Chair			
17	Table			

Part 6. Extension Services

1. Did anyone in your household visit an agricultural extension agent or an agricultural extension center during the last 12 months or did an agricultural extension agent visit your household?

1 – YES
0 – NO ☐

2. How many times during the last 12 months did members of your household discuss agricultural practices (growing crops / raising animals / planting trees) with such an agent?

☐ **QUANTITY**

3. How long did these meetings take on the average?

☐ **MINUTES**

4. Did you have to pay for the assistance of the agricultural extension agent or the agricultural extension center?

1 – YES
0 – NO ☐

5. How much did you pay for this assistance during the last 12 months?

☐ **Tsh**

6. What kind of service did you get from these visits?

1 - USE OF FERTILIZER
2 - IRRIGATION
3 - NEW SEED VARIETIES
4 - PEST INFESTATION
5 – CROP DISEASE
6 - SOIL PROBLEMS
7 - WEATHER PROBLEMS
8 - GENERAL CROP ADVICE
9 - VACCINATION SERVICES
10 - ANIMAL DISEASES
11 - ANIMAL FEED/NUTRITION
12 - INSEMINATION SERVICES
13 - MARKETING ADVICE
14 - HELP GETTING CREDIT
15 - FORESTRY RELATED ADVICE
16 - OTHER (SPECIFY____________________)

1st ☐ **2nd** ☐ **3rd** ☐

ALLOW UP TO THREE RESPONSES AND RANK THEM

Part 7. Housing

ENUMERATOR PLEASE OBSERVE AND NOTE THE FOLLOWING CONSTRUCTION MATERIALS:

1. WHAT IS THE MAJOR CONSTRUCTION MATERIAL OF THE EXTERNAL WALLS OF THE MAIN HOUSE?

1 – BURNED BRICK
2 - UNBURNED BRICK
3 - CONCRETE
4 - WOOD, LOGS
5 - TIN, ZINC SHEETING
6 – MUD AND/OR WOODEN POLES
7 – BAMBOO / STRAWS
8 - CANVAS, FELT
9 - OTHER (SPECIFY____________________) ☐

2. WHAT IS THE MAJOR MATERIAL OF THE ROOF OF THE MAIN HOUSE?

1 - METAL / ASBESTOS SHEETING
2 - TILES
3 – THATCH
4 – MUD (TEMBE)
5 - OTHER (SPECIFY ____________________) ☐

3. ENUMERATOR: PLS, ESTIMATE THE SQUARE METERS OF THE MAIN HOUSE

SQUARE METERS [ ]

4. How long has your household been living in this dwelling?

[ ] YEARS

5. What is the main source of water for drinking and cooking for your household in the rainy season?

1 - PRIVATE CONNECTION TO PIPELINE
2 - PRIVATE WELL
3 - PUBLIC TAPS/ STANDPIPE
4 - PUBLIC WELL/SPRING
5 – NEIGHBOR WELL
6 – NEIGHBOR PIPELINE
7 - WATER VENDOR
8 - RIVER, STREAM, LAKE, POND
9 - RAINWATER
10 - BOTTLED
11 - OTHER (SPECIFY ______________________)

6. During dry and rainy season, is the source of water for drinking and cooking for your household the same?

1 – YES
0 – NO

7. What is the main source of lighting in your dwelling ?

1 - ELECTRICITY
2 – KEROSENE HURRICANE LAMPS
3 – KEROSENE LAMP
4 – KEROSENE BURNER
5 – OPEN FIRE
6 - CANDLES OR BATTERY FLASHLIGHTS
7 - NO LIGHTING

8. What fuel do you use most often for cooking?

1 - GAS
2 - ELECTRICITY
3 - FIREWOOD
4 - CHARCOAL
5 - KEROSENE
6 – COW DUNG
7 - OTHER (SPECIFY ______________).

9. If you were to sell this dwelling today how much would you receive for it?

[ ] Tsh

Part 8. General Expenditures

1.How much did your family spend on the following items last year, either in a typical month (for example: in November) or in the entire year? Please give me an estimate.		
Item	**per month, Tsh**	**or total for one year Tsh**
1.Medicine / Health Costs		
2.Clothing and Shoes		
3.Festivities (ngoma, Idd, Christmas)		
4.Transport and Communication		
5.Rent of equipment / land		
6.External hired labor		
7.Other farm related expenditures such as seeds, tools etc.		
8.Other household consumables (soap, kerosene etc.)		
9.Education (School fees, Uniforms, Books etc.)		
10.Assistance to Relatives/Friends		
11.Costs of Funerals		
12.Other Expenditures (Specify)		

Part 9. Transport and Distances

The following questions refer to transport and distances from your home to different destinations

	1.How far from your household is...		2.How long takes the journey to [...] (THERE AND BACK)?	3.What means of transport do you choose to go there? 1 -CAR 2 -BIKE 3 -BUS 4 -MOTORCYCLE 5 -WALKING 6 -HAND CART 7 -OTHER (describe)	4.How often per month do you go there?	5.How much do you have to pay to go there and back?	6.Which are the conditions for access? 1 -ALWAYS TRAFFICABLE 2 -TRAFFICABLE ONLY DURING DRY SEASON 3 -OTHER (describe)
	Place	Distance (km)	Minutes	Code	Quantity	Tsh	Code
1.	village government office						
2.	the health center?						
3.	the school?						
4.	the church/mosque?						
5.	the market where you sell your products?						
6.	the shop where you buy agricultural inputs, such as fertilizer, seeds & farm equipment						

Part 10. Household Integration – Memberships

1.Do you or anybody else in your household belong to any trade union, cooperative, association, community council, church group or other group like a savings & credit cooperative society (SACCOS)?

1 – YES
0 – NO

	2.What are the names of **all** associations, cooperatives, trade unions, church or other groups where any member of your family takes part in?	3.Since when does he/she take part in…?	4.How many times per year does he/she go to assemblies of …?	5.How much does he/she have to pay per year to take part in …? IF HE/SHE DOESN'T PAY, SET "0"	6.Does this organization include only people of the same gender, ethnicity, religion or geographical region? **1 – YES** **0 – NO**	7.In this organization decisions are made in the following way: **1 – MEMBERS DISCUSS AND VOTE** **2 – DECISIONS ARE MADE BY LEADER**	8.In this organization the respective leader earns his/her livelihood in a different way than group members? **1 – YES** **0 – NO**
	Name	**Year**	**Quantity**	**Tsh**	**Coding**	**Coding**	**Coding**
1.							
2.							
3.							
4.							
5.							
6.							

Part 11. Off-Farm Income

With the following questions, we try to understand how important other, non-farm activities are, besides your agricultural income sources.

	1.Within the last year, did your family realize any income of the following kind?		2.Who of the family realized this income?	3.How much did the person(s) get per year?
	COMPLETE THIS COLUMN BEFORE GOING ON		WRITE DOWN THE ID(S) OF THE PERSON(S) (TABLE 1)	WRITE DOWN THE SUM PER YEAR FOR ALL PERSONS (USE YOR CALCULATOR)
		1 - YES **0 - NO**	**ID**	**Tsh per Year**
1.	Pensions/Retirement			
2.	Remittances/Help from family members who do not live here			
3.	Agricultural activities outside your lot			
4.	Wage employment/Other activities besides agriculture			
5.				

Part 12. Market and Price Information

1.Do you use the following sources for agricultural market and price information?

	Information source	1 - YES 0 - NO		Information source	1 - YES 0 - NO
1.	Newspaper		5.	Conversation with farmer leaders	
2.	Radio		6.	Conversation with friends/relatives	
3.	Telephone		7.	Trade union, cooperative, association, community council, church group	
4.	Conversation with extension staff		8.	Other (________________________)	

Part 13. Credit Availability

1.In your community are there any institutions from which you could get a loan to start a new agricultural business?

1 – YES
0 – NO ☐

	2.If you needed a loan of approximately 100 000 Tsh for starting an agricultural business, from what sources might you obtain it? ALLOW UP TO THREE RESPONSES **0 – CAN NOT GET ANY LOAN** **1 - GOVERNMENT BANK** **2 - PRIVATE BANK** **3 - SACCOS** **4 - COMMUNITY GROUP** **5 - FAMILY/FRIEND** **6 - OTHER (SPECIFY______)**	3.What interest rate might someone in the community expect to pay for a loan of this amount from this source?
		Percent
1.		
2.		
3.		

Part 14. TIST Small Group Membership

1.Are you participating in TIST – The International Small Group and Tree Planting Program?

1 – YES
0 – NO

IF NO: END OF INTERVIEW: PLEASE THANK THE INTERVIEWED PERSON FOR HIS/HER KIND COOPERATION

	Code	PLS FILL IN	Code	VERIFY INFORMATION WITH TIST MASTER
2.TIST Group Name			[gpname]	
3.TIST Group Number			[gpnumber]	
4.TIST Group Center			[cdgpcenter]	
5.TIST Group Node			[node]	
6.When did your Small Group actually join the TIST program?	Year		[regdate]	

7.Do you normally attend the Small Group meetings?	1 - YES 0 - NO			
8.How many farmers are members of the TIST Small Group you are belonging to?	Quantity		[members]	
9.When did you personally join the TIST Small Group?	Year			
10.How often did your Group meet last month?	Quantity		[datelastmeet]	
11.How many hours takes a normal meeting of your Group?	Hours		[lengthlastmeet]	
12.How many different leaders did your Group already have?	Quantity		[diffleaders]	
13.Does your Group have a bank account?	1 – YES 0 – NO		[cdacc]	
14.How many farmers do conservation farming in this Small Group?	Quantity		[cfmembers]	
15.Did your group ever submit *best practices* or suggest *program improvements* together with the sgmr	1 – YES 0 – NO		[bp / imp]	

16.Does the land on which the trees are planted belong to several different farmers?	1 – YES 0 – NO	
17.Do you share land for other activities besides the planting of trees?	1 – YES 0 – NO	
18.What is the main purpose of your tree planting?	1 – FIREWOOD 2 – FODDER 3 – TIMBER / POLES 4 – FRUITS 5 – MEDICINE 6 – OTHER DOMESTIC USES 7 – OTHER PRODUCTS FOR SALE 8 – VOUCHER PAYMENTS	
19.How many times have you been visited by a team of TIST quantifiers in the last year?	0 – NO VISIT 1 – ONE VISIT 2 – TWO VISITS	
20.Did you ever hear something about the "Kyoto Protocol"?	1 – YES 0 – NO	

21.What is the total land available for tree planting activities of the Small Group?	Quantity		1 -ACRES 2 -HECTARS 0 -OTHER (______________)		
22. On what kind of land are the trees planted?	1 – PRIVATELY OWNED 2 – COMMUNITY OWNED				
23.Do you help each other out in the Small Group while working on the fields?	1 – YES 0 – NO		24.When during the year is that usually?	1 – January 2 – February ... 12 – December	
25.Has this group currently more than 1000 alive trees?	1 – YES 0 – NO		[current or resting group?]		
26.Do you sometimes exchange food or crops amongst Small Group members?	1 – YES 0 – NO				
27.Did your Group ever receive a voucher for its activities?	1 – YES 0 – NO		[voucherdate]		
28.Was the money distributed to the Group members?			1 – YES 0 – NO		
29.Does every Group ember receive the same amount of money?			1 – YES 0 – NO		

30.According to which rule is the money distributed amongst Small Group members?	0 – EQUALLY 1 – LAND CONTRIBUTED TO GROUP 2 – LABOR CONTRIBUTED TO GROUP 3 – LEADERSHIP 4 – ACCORDING TO HOUSEHOLD SIZE / NUMBER OF ADULTS 5 – (______________________________)	

31.Which payment scheme do you prefer?	1 – ANNUALLY 2 – HALF-YEARLY 3 – EVERY THIRD MONTH 4 – EVERY SECOND MONTH 5 – MONTHLY	
32.Which contract length would you prefer for the planting of trees?	1 – FIVE YEAR CONTRACT (RENEWABLE) 2 – 20 YEAR CONTRACT (RENEWABLE) 3 – 30 YEAR CONTRACT 4 – 60 YEAR CONTRACT 5 – LONGER THEN 60 YEARS	

33.How did you hear about the TIST program?

	Information source	1 - YES 0 - NO
1.	Newspaper	
2.	Radio	
3.	Telephone	
4.	Conversation with TIST staff member or extension staff	
5.	Conversation with farmer leader	
6.	Conversation with friends/relatives/neighbors	
7.	Trade union, cooperative, association, community council, church group	
8.	Other (specify)	

ENUMERATOR: PLEASE MAKE SURE THAT YOU

- **OBSERVED** THE MAJOR **CONSTRUCTION MATERIAL OF THE** EXTERNAL **WALLS** OF THE HOUSE AND
- THE MAJOR **MATERIAL OF THE ROOF** IN PART 7 OF THE QUESTIONNAIRE

PLEASE **RECORD THE INTERVIEW TIME** ON PAGE 1

END OF INTERVIEW: THANK THE INTERVIEWED PERSON FOR HIS/HER KIND COOPERATION

EXTRA SHEET FOR **Part 2. Land Use**

Please draw all your lots/pieces of land and show all the different lots and plants, e.g. annual crops and permanent crops [mixed crops and monocultures], pastures, woods, (yard, small garden), house, (small rivers), and so on. Give each lot / Crop a number, mark the area actually under cultivation and the date, when the crop was planted.

App. 5 Community Questionnaire

The information provided herein will be kept confidential and will only be used for academic purposes

Project on Carbon Finance and Rural Development in Tanzania

Community Questionnaire

Center for Development Research, University of Bonn, Germany
Institute for Agricultural Policy and Market Research, Justus-Liebig-University, Giessen, Germany
TIST – The International Small Group and Tree Planting Project, Morogoro, Tanzania
Sokoine University of Agriculture, Morogoro, Tanzania

ALL FIELDS IN LIGHT GREY TO BE FILLED OUT BY ENUMERATOR

Part 0. Community Information

Questionnaire Number					
Recorder					
Date of Interview					
Time taken		Start		End	

Data Checked by	Date	Status		Comments
		OK	Return	

Data Entry	Name	Date	Comments
Pages			
Pages			

Region	
District	
Ward	
Village	
Altitude	

Composition of the Focus Group

Respondent Status	Surname	Name	Gender

1.How large is the total land area of this community (government land and privately owned land)?

[ ] HECTARES

2.How large is the total land area that is owned by the government in this community?

[ ] HECTARES

3.How large is the total land area that is privately owned in this community?

[ ] HECTARES

4.How much of the land owned by the government has been allocated to village members?

[ ] HECTARES

5.How much land is left as a reserve?

[ ] HECTARES

6.The allocation of government land in the past was mainly for ...

	Forestry
	Mining
	Grazing
	Other (Specify______________________________)

7.The government land kept as a reserve will mainly be allocated to ...

	Forestry
	Mining
	Grazing
	Other (Specify______________________________)

8.How many people live in this community?

[ ] NUMBER OF PEOPLE

9.How many households are found in this community?

[ ] NUMBER OF HOUSEHOLDS

10.Are you aware of the TIST program, which stands for the "International Small Group and Tree Planting Program"?

[ ] 1 – YES
0 – NO

11.What is that project doing?

____
____
____
____
____
____

12.How many households of this community are taking part in TIST - the International Small Group and Tree Planting Program (if so)?

[ ] NUMBER OF HOUSEHOLDS

13.	What is the price per liter of [...] in this community?
Fuel	Tsh
Petrol	
Diesel	
Kerosene	

14.	What is the price for [...] in this community?	1 - BAG 2 - TIN 3 - SACK 4 - BUNDLE 5 – CARTLOAD 6 – PIECE 7 - OTHER (SPECIFY______).	
		Quantity	Tsh
1.	Charcoal		
2.	Manure		
3.	Firewood		
4.	Poles	1piece 2x6 inch x12 feet	

15.	If someone in this community needed a loan of approximately 100 000 Tsh for starting a business, from what sources might he or she obtain it? ALLOW UP TO THREE RESPONSES 0 – CAN NOT GET ANY LOAN 1 - GOVERNMENT BANK 2 - PRIVATE BANK 3 - SACCOS 4 - COMMUNITY GROUP 5 - FAMILY/FRIEND 6 - OTHER (SPECIFY_________)	What interest rate might someone in the community expect to pay for a loan of this amount from this source?
1.		
2.		
3.		

16.	Which activities are the most important sources of employment for individuals in this community? PLEASE RANK 1 - FARMING, FORESTRY, HUNTING, FISHING 2 - MINING 3 - INDUSTRY 4 – BUILDING, CONSTRUCTION 5 - TRANSPORTATION 6 - OTHER SERVICES (HAIRCUTTING, TAILORS, ETC) 7 - CRAFTS, SMALL-SCALE TRADE 8 – TOURISM 9 - OTHER (SPECIFY_________)	What is the average daily wage of an unskilled adult male?	What is the average daily wage of an unskilled adult female?	What is the average daily wage of a non-adult?
		Tsh	Tsh	Tsh
1st				
2nd				
3rd				

17.	Please list the three most important cash and food crops cultivated by residents of this community, in order of importance. **1 – RICE** **2 – BEANS** **3 – SORGHUM** **4 – MILLET** **5 – PIGEON PEAS** **6 – MAIZE** **7 – CASSAVA** **8 – SWEET POTATOES** **9 – VEGETABLES** **10 – COTTON** **11 – OTHER (SPECIFY____________)**	How many times is this crop harvested each year?	In which months is […] generally planted? **1 – January** **2 – February** … **12 – December** **IF NO PARTICULAR MONTH CODE AS 13.**			In which months is […] generally harvested? **1 – January** **2 – February** … **12 – December** **IF NO PARTICULAR MONTH CODE AS 13.**			In general to whom is the harvest of […] sold? **1 - LOCAL MARKET** **2 - GOVERNMENT** **3 - COOPERATIVE** **4 - PRIVATE BUYER** **5 - NOT SOLD** **6 - OTHER(SPECIFY_)**		
		Times	**1ST**	**2ND**	**3RD**	**1ST**	**2ND**	**3RD**	**1ST**	**2ND**	**3RD**
FOOD	A										
	B										
	C										
CASH	A										
	B										
	C										

18.	What is the price of […] in this community?	Tsh	**1 – Kilogram** **2 - Liter** **Code**
1.	Artificial Fertilizer		
2.	Insecticide #1		
3.	Insecticide #2		
4.	Seeds		
5.	Tractor ploughing		per acre

19.Do you think that the life of the people in this community is better or worse than it was ten years ago?

1 - BETTER
2 - WORSE
3 - NO CHANGE

20.Why?

ALLOW UP TP THREE RESPONSES

1 - CHANGES IN AGRICULTURAL POLICIES **1st**
2 - EXPANSION IN NON-AGRICULTURAL EMPLOYMENT
3 - WEATHER
4 - INFLATION
5 - CHANGES IN CONDITIONS FOR IMPROVING SKILLS **2nd**
6 - CHANGES IN ACCESSIBILITY OF MEDICAL CARE
7 - CHANGES IN ACCESSIBILITY OF SOCIAL SERVICES
8 - NATURAL DISASTER, DROUGHT
9 - INSECT INFESTATION
10 - CROP DISEASE **3rd**
11 - OTHER (SPECIFY________________________________)

21.In the last five years, have there been more people who moved into this community or more people who moved away?

1 - MORE MOVED IN
2 - MORE MOVED AWAY
3 - ABOUT THE SAME OF BOTH
4 - NEITHER ARRIVALS NOR DEPARTURES

22.Who is most likely to move out of this community?

	Young Males
	Young Females
	Educated People
	Non-indigenous People
	Other (____)

23.Whom do you consider to be "poor" in this community?

____
____
____
____
____

24.In thinking about the poorest residents of your community, what factors contribute to their situation?

ALLOW UP TP THREE RESPONSES

1 - LACK OF EMPLOYMENT OPPORTUNITIES **1st**
2 - POOR HEALTH
3 - LOW LEVEL OF EDUCATION AND SKILLS
4 - POOR LAND PRODUCTIVITY
5 - NO WAY TO BORROW MONEY **2nd**
6 - TO OLD TO WORK
7 - LAZY
8 - BAD LUCK
9 - OTHER (SPECIFY_____) **3rd**

25.Now think about the weather conditions in the past 12 months. How does [....] this year compare to what you consider to be normal conditions? That is, is it much higher, somewhat higher, about normal, somewhat lower, or much lower compared to normal conditions?

1 - MUCH MORE/HIGHER THAN USUAL …rainfall
2 - SOMEWHAT MORE/HIGHER THAN USUAL
3 - ABOUT THE SAME
4 - SOMEWHAT LESS THAN USUAL
5 - MUCH LESS THAN USUAL
…temperature

26.			STANDARDIZATION			
FOOD CODE	In what unit is [...] usually purchased in this community? **1 – BAG** **2 - TIN** **3 - SACK** **4 - BUNDLE** **5 - CAN** **6 - KILOGRAM** **7 - PIECE** **8 - OTHER (SPECIFY____________)**		How much [...] is contained in one [UNIT FROM QUESTION 1]? **1 – KILOGRAM** **2 - GRAM** **3 - LITER** **4 - CC** **5 - OTHER (SPECIFY__________)**		What is the price for one [UNIT FROM QUESTION 1] of [....] during harvesting season?	What is the price for one [UNIT FROM QUESTION 1] of [....] during season of shortage?
	Item	**Unit**	**Amount**	**Unit Code**	**Tsh**	**Tsh**
1	Maize					
2	Paddy					
3	Sorghum					
4	Bulrush Millet					
5	Finger Millet					
6	Cassava					
7	Sweet Potatoes					
8	Onions					
9	Beans					
10	Cowpeas					
11	Pigeon Peas					
12	Bambara Nuts					
13	Sunflower					
14	Simsim					
15	Groundnut					
16	Caster Seed					
17	Cotton					
18	Cabbage					
19	Tomatoes					
20	Spinach					
21	Carrot					
22	Amaranths					
23	Pumpkins					
24	Cucumber					
25	Egg Plant					
26	Water Melon					
27	Sugar Cane					
28	Banana					
29	Mango					
30	Paw paw					
31	Orange					
32	Grapes					
33	Guava					
34	Lime / Lemon					
35	Beef					
36	Pork					

37	Chicken					
38	Fish					
39	Egg					
40	Milk					
41	Beer					
42	Cigarette					
43	Kanga / Kitenge					
44	Shirt / Dress					
45	Trousers					
46	Shoes					
47	Salt					
48	Sugar					
49	Cooking Oil					

References

Adams, W. M., Aveling, R., Brockington, D., Dickson, B., Elliott, J., Hutton, J., Roe, D., Vira, B., and Wolmer, W. (2004): "Biodiversity conservation and the eradication of poverty." Science, 306(5699), 1146-1149.

Adesina, A. A. and Chianu, J. (2002): "Determinants of farmers' adoption and adaptation of alley farming technology in Nigeria." Agroforestry Systems, 55(2), 99-112.

Ajayi, O. C., Franzel, S., Kuntashula, E., and Kwesiga, F. (2003): "Adoption of improved fallow technology for soil fertility management in Zambia: Empirical studies and emerging issues." Agroforestry Systems, 59(3), 317-326.

Akaike, H. (1987): "Factor-Analysis and Aic." Psychometrika, 52(3), 317-332.

Angelsen, A. (2005): Personal communication with Arild Angelsen, Dept. of Economics and Resource Management, Norwegian University of Life Sciences.

Angelsen, A. and Wunder, S. (2003): "Exploring the Forest—Poverty Link: Key Concepts, Issues and Research Implications." Rep. No. CIFOR Occasional Paper No. 40 , Center for International Forestry Research - CIFOR, Bogor, Indonesia.

Arbuckle, J. L. and Wothke, W. (1999): AMOS 4.0 User`s Guide, SmallWaters Corporation, Chicago.

Arnold, J. E. M. (1983): "Economic considerations in agroforestry projects." Agroforestry Systems, 1(4), 299-311.

Atta-Krah, A. N. and Francis, P. A. (1987): "The role of on-farm trials in the evaluation of composite technologies: The case of alley farming in Southern Nigeria." Agricultural Systems, 23, 133-152.

Ayuk, E. T. (1997): "Adoption of agroforestry technology: The case of live hedges in the central plateau of Burkina Faso." Agricultural Systems, 54(2), 189-206.

Backhaus, K., Erichson, B., Plinke, W., and Weiber, R. (2006): Multivariate Analysemethoden. Eine anwendungsorientierte Einführung. 11.Auflage, Springer, Berlin, Heidelberg, New York.

Baker, J. L. (2000): Evaluating the Impact of Development Projects on Poverty - A Handbook for Practitioners, The World Bank, Washington, DC.

Bandalos, D. L. (1993): "Factors Influencing Cross-Validation of Confirmatory Factor-Analysis Models." Multivariate Behavioral Research, 28(3), 351-374.

Barrett, C. B., Lee, D. R., and Mcpeak, J. G. (2005): "Institutional arrangements for rural poverty reduction and resource conservation." World Development, 33(2), 193-197.

Bass, S., Dubois, O., Moura Costa, P., Pinard, M., Tipper, R., and Wilson, C. (2000): Rural livelihoods and carbon management, IIED - International Institute for Environment and Development, London, UK.

Behrmann, J. R. and Raylynn, O. (2000): "Basic Economic Models and Econometric Tools." Designing Household Survey Questionnaires for Developing Countries, M. Grosh and P. Glewwe, eds., The World Bank, Washington, D.C., 365-389.

Bekele, W. and Drake, L. (2003): "Soil and water conservation decision behavior of subsistence farmers in the Eastern Highlands of Ethiopia: a case study of the Hunde-Lafto area." Ecological Economics, 46(3), 437-451.

Bentler, P. M. (1985): Theory and Implementation of EQS: A Structural Equations Program, Los Angeles.

Bentler, P. M. (1990): "Comparative Fit Indexes in Structural Models." Psychological Bulletin, 107(2), 238-246.

Bentler, P. M. (1992): "On the Fit of Models to Covariances and Methodology to the Bulletin." Psychological Bulletin, 112(3), 400-404.

Bentler, P. M. and Bonett, D. G. (1980): "Significance Tests and Goodness of Fit in the Analysis of Covariance-Structures." Psychological Bulletin, 88(3), 588-606.

Bentler, P. M. and Wu, E. J. C. (1995): EQS for Windows: User`s Guide, Enico CA: Multivariate Software, Inc..

BioCF (2004): Carbon Finance Document: Tanzania Small Group and Tree Planting Program, The World Bank, Carbon Finance Business, Washington DC, USA.

BioCF (2006): BioCarbon Fund Project Portfolio. http://carbonfinance.org/Router.cfm?Page=BioCF&FID=9708&ItemID=9708&ft=Projects&ProjID=9642&stp=Yes. Accessed: 29. 09. 2006 .

Bollen, K. A. (1989): Structural Equations with Latent Variables, John Wiley & Sons, New York.

Bozdogan, H. (1987): "Model Selection and Akaike Information Criterion (Aic) - the General-Theory and Its Analytical Extensions." Psychometrika, 52(3), 345-370.

Browne, M. W. and Cudeck, R. (1989): "Single sample cross-validation indices for covariance structures." Multivariate Behavioral Research, 24, 445-455.

Browne, M. W. and Cudeck, R. (1993): "Alternative ways of assessing model fit." Testing structural equation models, K. A. Bollen and J. S. Long, eds., Sage, Newbury Park, CA, 445-455.

Byrne, B. M. (2001): Structural Equation Modeling With Amos. Basic Concepts, Applications and Programming, Lawrence Erlbaum Associates, Mahwah, New Jersey.

Cacho, O. J., Marshall, G. R., and Milne, M. (2003): "Smallholder Agroforestry Projects: Potential for Carbon Sequestration and Poverty Alleviation." Rep. No. ESA Working Paper No. 03-06.

Carney, D. (2002): "Sustainable Livelihood Approaches: Progress and Possibilities for Change." D.

Castello, C., Stearns, K., and Christen, R. P. (1991): Exposing Interest Rates: Their True Significance to Microentrepreneurs and Credit Programs , Cambridge, Mass. USA.

Cavendish, W. (2002): "Quantitative methods for estimating the economic value of resource use to rural households." Uncovering the Hidden Harvest, B. M. Campbell and M. K. Luckert, eds., Earthscan Publications, London, 17-65.

Caviglia-Harris, J. L. (2003): "Sustainable agricultural practices in Rondonia, Brazil: Do local farmer organizations affect adoption rates?" Economic Development and Cultural Change, 52(1), 23-49.

Chambers, R. and Conway, G. R. (1991): Sustainable rural livelihoods: practical concepts for the 21st century, IDS - Institute of Development Studies, Sussex, UK.

Chianu, J. N. and Tsujii, H. (2004): "Determinants of farmers' decision to adopt or not adopt inorganic fertilizer in the savannas of northern Nigeria." Nutrient Cycling in Agroecosystems, 70(3), 293-301.

Christen, R. P., Rhyne, E., Vogel, R. C., and McKean, C. (1995): Maximizing the Outreach of Microenterprise Finance: An Analysis of Successful Microfinance Programs , Washington DC, USA.

Conley, T. and Udry, C. (2001): "Social learning through networks: The adoption of new agricultural technologies in Ghana." American Journal of Agricultural Economics, 83(3), 668-673.

Current, D., Lutz, E., and Scherr, S. (1995): Costs, Benefits, and Farmer Adoption of Agroforestry: Project Experience in Central America and the Carribean, The World Bank, Washington D.C.

Dagang, A. B. K. and Nair, P. K. R. (2003): "Silvopastoral research and adoption in Central America: recent findings and recommendations for future directions." Agroforestry Systems, 59(2), 149-155.

Daly, F., Hand, D. J., Jones, M. C., Lunn, A. D., and McConway, K. J. (1995): Elements of Statistics, The Open University/Addison-Wesley, Harlow.

Deaton, A. and Grosh, M. (2000): "Consumption." Designing Household Survey Questionnaires for Developing Countries. Lessons from 15 years of the Living Standards Measurement Study, M. Grosh and P. Glewwe, eds., The World Bank, Washington D.C., USA.

DeFries, R. S., Houghton, R. A., Hansen, M. C., Field, C. B., Skole, D., and Townshend, J. (2002): "Carbon emissions from tropical deforestation and regrowth based on satellite observations for the 1980s and 1990s." Proceedings of the National Academy of Sciences of the United States of America, 99(22), 14256-14261.

Delgado, C. L., Hopkins, J., and Kelly, V. A. (1998): "Agricultural Growth Linkages in Sub-Saharan Africa." Rep. No. 107, IFPRI, Washington D.C.

Dixon, J., Gulliver, A., and Gibbon, D. (2001): Farming Systems and Poverty, Food and Agriculture Organization (FAO) and The World Bank, Rome, Italy and Washington D.C., USA.

Echevarría, M. (2002): "Financing watershed conservation: The FONAG water fund in Quito, Ecuador." Selling Forest Environmental Services. Market-based Mechanisms for Conservation and Development, S. Pagiola, J. Bishop, and N. Landell-Mills, eds., Earthscan, London, UK, 91-101.

Echevarría, M., Vogel, J., Albán, M., and Meneses, F. (2003): "The impacts of payments for watershed services in Ecuador. Emerging lessons from Pimampiro and Cuenca." IIED - International Institute for Environment and Development, London, UK.

Fan, X., Thompson, B., and Wang, L. (1999): "Effects of sample size, estimation methods, and model specification on structural equation modeling fit indices." Structural Equation Modeling: A Multidisciplinary Journal, 6, 56-83.

FAO (1997): Conservation Farming Handbook. For Small Holders in Regions I & II, FAO - Food and Agriculture Organization of the United Nations, Conservation Farming Unit, Rome, Italy.

FAO (2000): Global Forest Resources Assessment 2000 - Main Report. Rome, Italy, FAO - Food and Agriculture Organization of the United Nations.

Fearnside, P. M. (2001): "Saving tropical forests as a global warming countermeasure: an issue that divides the environmental movement." Ecological Economics, 39, 167-184.

Feder, G., Just, R. E., and Zilberman, D. (1985): "Adoption of Agricultural Innovations in Developing Countries: A Survey." Economic Development and Cultural Change, 33(2), 255-298.

Ferraro, P. J. and Simpson, R. D. (2005): "Cost-effective conservation when eco-entrepreneurs have market power." Environment and Development Economics, 10, 651-663.

Frank, S. (2005): Personal communication with Sikitu Frank, TIST program manager, during field research in March 2005.

Fuwape, J. A. and Akindele, S. O. (1997): "Biomass yield and energy value of some fast-growing multipurpose trees in Nigeria." Biomass & Bioenergy, 12(2), 101-106.

Ghadim, A. K. A. and Pannell, D. J. (1999): "A conceptual framework of adoption of an agricultural innovation." Agricultural Economics, 21(2), 145-154.

Glendinning, A., Mahapatra, A., and Mitchell, C. P. (2001): "Modes of communication and effectiveness of agroforestry extension in eastern India." Human Ecology, 29(3), 283-305.

Glewwe, P. (2000): "Household Roster." Designing Household Survey Questionnaires for Developing Countries. Lessons from 15 years of the Living Standards Measurement Study, M. Grosh and P. Glewwe, eds., The World Bank, Washington D.C., USA.

Goldberger, A. S. (1972): "Structural equation methods in the social sciences." Econometrica, 40, 979-1001.

Grieg-Gran, M., Porras, I., and Wunder, S. (2005): "How can market mechanisms for forest environmental services help the poor? Preliminary lessons from Latin America." World Development, 33(9), 1511-1527.

Grootaert, C. (1982): "The Conceptual Basis of Measures of Household Welfare and their Implied Survey Data Requirements." Rep. No. 19, The World Bank.

Grosh, M. and Glewwe, P. (2000): "Making Decisions on the Overall Design of the Survey." Designing Household Questionnaires for Developing Countries. Lessons from 15 Years of the Living Standards Measurement Study, M. Grosh and P. Glewwe, eds., The World Bank, Washington D.C., USA, 21-41.

Gutierrez, V. H., Zapata, M., Sierra, C., Laguado, W., and Santacruz, A. (2006): "Maximizing the profitability of forestry projects under the Clean Development Mechanism using a forest management optimization model." Forest Ecology and Management, 226(1-3), 341-350.

Hamilton, K., Bayon, R., Turner, G., and Higgins, D. (2007): State of the Voluntary Carbon Market 2007. Picking Up Steam, Washington DC, USA.

Hardner, J. J., Frumhoff, P. C., and Goetze, D. C. (2000): "Prospects for mitigating carbon, conserving biodiversity, and promoting socioeconomic development objectives through the clean development mechanism." Mitigation and Adaptation Strategies for Global Change, V5(1), 61-80.

Helms, B. and Reille, X. (2004): Interest rate ceilings and microfinance: the story so far, CGAP - Consultative Group to Assist the Poor, Washington DC, USA.

Herath, P. H. M. U. and Takeya, H. (2003): "Factors determining intercropping by rubber smallholders in Sri Lanka: a logit analysis." Agricultural Economics, 29(2), 159-168.

Hines, D. A. and Eckman, K. (1993): Indigenous multipurpose trees of Tanzania: Uses and economic benefits for people, Cultural Survival Canada and Development Services Foundation of Tanzania, Ottawa, Ontario, Canada.

Hoelter, J. W. (1983): "The Analysis of Covariance-Structures - Goodness-Of-Fit Indexes." Sociological Methods & Research, 11(3), 325-344.

Houghton, R. A. (2003): "Revised estimates of the annual net flux of carbon to the atmosphere from changes in land use and land management 1850-2000." Tellus Series B-Chemical and Physical Meteorology, 55(2), 378-390.

Hoyle, R. H. (1995): Structural Equation Modeling - Concepts, Issues, and Applications, SAGE Publications, Thousand Oaks, London, New Delhi.

Hu, L.-T. and Bentler, P. M. (1995): "Evaluating model fit." Structural equation modeling: Concepts, issues, and applications, R. H. Hoyle, ed., Sage, Thousand Oaks, CA, 76-99.

Hu, L.-T. and Bentler, P. M. (1999): "Cutoff criteria for fit indexes in covariance structure analysis: Conventional criteria versus new alternatives." Structural Equation Modeling: A Multidisciplinary Journal, 6, 1-55.

ICRAF (2006): World Agroforestry Centre - Agroforestree Database. http://www.worldagroforestrycentre. org/Sites/TreeDBS/aft. asp, accessed 15.12.2006.

IIRR and ACTN (2005): Conservation agriculture. A manual for farmers and extension workers in Africa, IIRR - International Institute of Rural Reconstruction, Nairobi, Kenya and ACTN - African Conservation Tillage Network, Harare, Zimbabwe.

International Co-operative Alliance (2001): "Rural savings and credit co-operative societies in Tanzania." International Co-operative Alliance Regional Office for East and Southern Africa, Nairobi, Kenya.

IPCC (2001): "Climate Change 2001: The Scientific Basis - Contribution of Working Group I to the Third Assessment Report of the Intergovernmental Panel on Climate Change." J. T. Houghton, Y. Ding, D. J. Griggs, M. Noguer, P. J. Van der Linden, X. Dai, K. Maskell, and C. A. Johnson, eds., Cambridge University Press, Cambridge, UK.

IPP Media (2005): SACCOS in dire need of trained personnel. IPP Media Online .Dar Es Salaam, Tanzania, IPP Media, http://www.ippmedia.com/ipp/guardian/2005/11/04 /53357.html.

Isham, J. (2002): "The effect of social capital on fertiliser adoption: Evidence from rural Tanzania." Journal of African Economies, 11(1), 39-60.

ISI (2006): ISI Web of Science. http://portal.isiknowledge.com/portal.cgi?DestApp=WOS& Func=Frame, accessed 05.04.2006.

James, L. R., Mulaik, S. A., and Brett, J. M. (1982): Causal analysis: Assumptions, models, and data, Sage, Beverly Hills, CA.

Jones, D. W. and Oneill, R. V. (1992): "Land-Use with Endogenous Environmental Degradation and Conservation." Resources and Energy, 14(4), 381-400.

Jöreskog, K. G. (1993): "Testing structural equation models." Testing structural equation models, K. A. Bollen and J. S. Long, eds., Sage, Newbury Park, CA, 294-316.

Jöreskog, K. G. and Sörbom, D. (1984): LISREL - VI user`s guide (3rd. ed.), Scientific Software, Mooresville, IN.

Jöreskog, K. G. and Sörbom, D. (1988): LISREL 7: A guide to the program and applications., SPSS, Chicago.

Jöreskog, K. G. and Sörbom, D. (1989): LISREL 7 - User`s Reference Guide, Mooresville.

Jöreskog, K. G. and Sörbom, D. (1996): LISREL 8: Structural equation modeling with the SIMPLIS command language, Scientific Software International, Chicago.

Kaimowitz, D. and Angelsen, A. (1998): Economic Models of Tropical Deforestation. A Review, CIFOR - Center for International Forestry Research, Bogor, Indonesia.

Kaliba, A. R. M., Verkuijl, H., Mwangi, W., Moshi, A. J., Chilagane, A., Kaswende, J. S., and Anandajayasekeram, P. (1998): "Adoption of Maize Production Technologies in Eastern Tanzania." International Maize and Wheat Improvement Center (CIMMYT), the United Republic of Tanzania, and the Southern Africa Centre for Cooperation in Agricultural Research (SACCAR), Mexico.

Kebede, Y. (1992): "Risk behavior and new agricultural technologies: the case of producers in the central highlands of Ethiopia ." Quarterly Journal of International Agriculture, 31, 269-284.

Kuntashula, E. and Mafongoya, P. L. (2005): "Farmer participatory evaluation of agroforestry trees in eastern Zambia." Agricultural Systems, 84(1), 39-53.

Landell-Mills, N. (2002): "Developing markets for forest environmental services: an opportunity for promoting equity while securing efficiency?" Carbon, biodiversity, conservation and income: an analysis of a free-market approach to land-use change and forestry in developing and developed countries, I. R. Swingland, E. C. Bettelheim, J. Grace, G. T. Prance, and L. S. Saunders, eds., The Royal Society, London, 1817-1825.

Landell-Mills, N. and Porras, I. T. (2002): "Silver bullet or fools' gold? A global review of markets for forest environmental services and their impact on the poor ." International Institute for Environment and Development - IIED, London.

Lapar, M. L. A. and Pandey, S. (1999): "Adoption of soil conservation: the case of the Philippine uplands." Agricultural Economics, 21(3), 241-256.

Leuschner, W. A. and Khaleque, K. (1987): "Homestead Agroforestry in Bangladesh." Agroforestry Systems, 5(2), 139-151.

Lipper, L. and Cavatassi, R. (2003): Land Use Change, Carbon Sequestration and Poverty Alleviation, ESA Working Paper No. 03-13, FAO - Food and Agriculture Organization, Rome, Italy.

Loehlin, J. C. (2004). Latent Variable Models - An Introduction to Factor, Path and Structural Equation Analysis. Fourth Edition, Lawrence Erlbaum Associates, Inc. Publishers, Mahwah, New Jersey.

Long, J. S. (1983): Confirmatory Factor Analysis: A Preface to LISREL, Beverly Hills u.a..

Los, S. O., Weedon, G. P., North, P. R. J., Kaduk, J. D., Taylor, C. M., and Cox, P. M. (2006): "An observation-based estimate of the strength of rainfall-vegetation interactions in the Sahel." Geophysical Research Letters, 33(16).

MacCallum, R. C., Browne, M. W., and Sugawara, H. M. (1996): "Power analysis and determination of sample size for covariance structure modeling." Psychological Methods, 1(2), 130-149.

Mahapatra, A. K. and Mitchell, C. P. (2001): "Classifying tree planters and non planters in a subsistence farming system using a discriminant analytical approach." Agroforestry Systems, 52(1), 41-52.

Marra, M., Pannell, D. J., and Ghadim, A. A. (2003): "The economics of risk, uncertainty and learning in the adoption of new agricultural technologies: where are we on the learning curve?" Agricultural Systems, 75(2-3), 215-234.

Mas-Colell, A., Whinston, M. D., and Green, J. R. (1995): Microeconomic Theory, Oxford University Press, New York.

Masudi, A., Ishumi, A., Mbeo, F., and Sambo, W. (2001): "Tanzania - Child Labour in Commercial Agriculture - Tobacco: A Rapid Assessment." International Labour Organization International Programme on the Elimination of Child Labour (IPEC) , Geneva.

Mbaga-Semgalawe, Z. and Folmer, H. (2000): "Household adoption behaviour of improved soil conservation: the case of the North Pare and West Usambara Mountains of Tanzania." Land Use Policy, 17(4), 321-336.

Mbuya, L. P., Msanga, H. P., Ruffo, C. K., Birnie, A., and Tengnäs, B. (1994): Useful trees and shrubs for Tanzania: identification, propagation and management for agricultural and pastoral communities, Regional Soil Conservation Unit (RSCU), Swedish International Development Authority (SIDA), Nairobi, Kenya.

Mellor, J. W. (1966): The Economics of Agricultural Development, Cornell University Press, Ithaca, USA.

Mercer, D. E. and Miller, R. P. (1998): "Socioeconomic research in agroforestry: Progress, prospects, priorities." Agroforestry Systems, 38(1-3), 177-193.

Mercer, D. E. and Pattanayak, S. K. (2003): "Agroforestry Adoption by Smallholders." Forests in a Market Economy, E. Sills and K. L. Abt, eds., Kluwer Academic Publishers, Dordrecht, Boston, London, 283-299.

Min.of Agriculture / NBS (2001): "District Intergraded Agricultural Survey 1998/99 - National Report." Statistics Unit - Ministry of Agriculture and Cooperatives / National Bureau of Statistics, Dar Es Salaam, Tanzania.

Molua, E. L. (2005): "The economics of tropical agroforestry systems: the case of agroforestry farms in Cameroon." Forest Policy and Economics, 7(2), 199-211.

Morris, S. S., Carletto, C., Hoddinott, J., and Christiaensen, L. J. M. (2000): "Validity of rapid estimates of household wealth and income for health surveys in rural Africa." Journal of Epidemiology and Community Health, 54(5), 381-387.

Moser, C. M. and Barrett, C. B. (2003): "The disappointing adoption dynamics of a yield-increasing, low external-input technology: the case of SRI in Madagascar." Agricultural Systems, 76(3), 1085-1100.

Mowo, J. G., Janssen, B. H., Oenema, O., German, L. A., Mrema, J. P., and Shemdoe, R. S. (2006): "Soil fertility evaluation and management by smallholder farmer communities in northern Tanzania." Agriculture Ecosystems & Environment, 116(1-2), 47-59.

Moyo, S. and Veeman, M. (2004): "Analysis of joint and endogenous technology choice for protein supplementation by smallholder dairy farmers in Zimbabwe." Agroforestry Systems, 60(3), 199-209.

Mulaik, S. A., James, L. R., Vanalstine, J., Bennett, N., Lind, S., and Stilwell, C. D. (1989): "Evaluation of Goodness-Of-Fit Indexes for Structural Equation Models." Psychological Bulletin, 105(3), 430-455.

Mutesasira, L., Sampangi, H., Mugwanga, H., Kashangaki, J., Maximambali, F., Lwoga, C., Hulme, D., Wright, G., and Rutherford, S. (1999): "Savings and the Poor the methods, use and impact of savings by the poor of East Africa." MicroSave-Africa, Kampala, Uganda.

Nair, P. K. R. (1996): "Agroforestry directions and trends." The Literature of Forestry and Agroforestry, P. McDonald and J. Lassoie, eds., Cornell University Press, Ithaca, NY, 74-95.

Narayan, D. (1997): Voices of the Poor. Poverty and Social Capital in Tanzania, The World Bank, Washington DC, USA.

Narayan, D. and Pritchett, L. (1999): "Cents and Sociability: Household Income and Social Capital in Rural Tanzania." Economic Development and Cultural Change, 871-897.

Neupane, R. P., Sharma, K. R., and Thapa, G. B. (2002): "Adoption of agroforestry in the hills of Nepal: a logistic regression analysis." Agricultural Systems, 72(3), 177-196.

Niles, J. O., Brown, S., Pretty, J., Ball, A. S., and Fay, J. (2002): "Potential carbon mitigation and income in developing countries from changes in use and management of agricultural and forest lands." Carbon, biodiversity, conservation and income: an analysis of a free-market approach to land-use change and forestry in developing and developed countries, I. R. Swingland, E. C. Bettelheim, J. Grace, G. T. Prance, and L. S. Saunders, eds., The Royal Society, London, 1621-1639.

Nowak, J., Blount, A., Workman, S. (2007): Circular 1430: Integrated Timber, Forage and Livestock Production - Benefits of Silvopasture. http://edis.ifas.ufl.edu/pdffiles/FR/FR13900.pdf - visited 20. 02. 2007 ., University of Florida - Institute of Food and Agricultural Sciences.

Oberthür, S. and Ott, H. E. (1999): The Kyoto Protocol. International Climate Policy for the 21st Century, Springer, Berlin; Heidelberg; New York.

Pagiola, S., Arcenas, A., and Platais, G. (2005): "Can payments for environmental services help reduce poverty? An exploration of the issues and the evidence to date from Latin America." World Development, 33(2), 237-253.

Pagiola, S., Landell-Mills, N., and Bishop, J. (2002): "Making market-based mechanisms work for forests and people." Selling Forest Environmental Services. Market-based Mechanisms for Conservation and Development, S. Pagiola, J. Bishop, and N. Landell-Mills, eds., Earthscan, London, UK, 261-289.

Pagiola, S., Ramirez, E., Gobbi, J., de Haan, C., Ibrahim, M., Murgueitio, E., and Ruiz, J. P. (2007): "Paying for the environmental services of silvopastoral practices in Nicaragua." Ecological Economics, 64(2), 374-385.

Pannell, D. J., Marshall, G. R., Barr, N., Curtis, A., Vanclay, F., and Wilkinson, R. (2006): "Understanding and promoting adoption of conservation practices by rural landholders." Australian Journal of Experimental Agriculture, 46(11), 1407-1424.

Pattanayak, S. K., Mercer, D. E., Sills, E., and Yang, J. C. (2003): "Taking stock of agroforestry adoption studies." Agroforestry Systems, 57(3), 137-150.

Pennings, J. M. E. and Smidts, A. (2003): "The shape of utility functions and organizational behavior." Management Science, 49(9), 1251-1263.

Poate, C. D. and Daplyn, P. F. (1993): Data for Agrarian Development, Cambridge University Press, Cambridge.

Pratt, D. J., Greenway, P. J., and Gwynne, M. D. (1966): "A Classification of East African Rangeland - with An Appendix on Terminology." Journal of Applied Ecology, 3(2), 369-382.

Puri, S. and Nair, P. K. R. (2004): "Agroforestry research for development in India: 25 years of experiences of a national program." Agroforestry Systems, 61(1), 437-452.

Putnam, R. (1995): "Bowling Alone: America's Declining Social Capital." Journal of Democracy, 6(1), 65-78.

Reardon, T. and Vosti, S. A. (1995): "Links between rural poverty and the environment in developing countries: Asset categories and investment poverty." World Development, 23(9), 1495-1506.

Reinecke, J. (2005): Strukturgleichungsmodelle in den Sozialwissenschaften, Oldenburg Wissenschaftsverlag GmbH, München.

Richards, K. and Andersson, K. (2001): "The leaky sink: persistent obstacles to a forest carbon sequestration program based on individual projects." Climate Policy, 1(1), 41-54.

Robinson, M. (2001): The Microfinance Revolution. Vol.1, Sustainable Finance for the Poor, Washington DC, USA.

Rogers, E. M. and Shoemaker, F. F. (1971): Communication of Innovations, Free Press, New York.

Rojas, M. and Aylward, B. (2002): The case of La Esperanza: a small, private, hydropower producer and a conservation NGO in Costa Rica. Land-water linkages in rural watersheds case study series, FAO - Food and Agriculture Organization, Rome, Italy.

Ruthenberg, H. (1980): Farming systems in the tropics, Clarendon Press, Oxford, UK.

Sadoulet, E. and De Janvry, A. (1995): Quantitative Development Policy Analysis, The Johns Hopkins University Press, Baltimore, Maryland.

Salam, M. A., Noguchi, T., and Koike, M. (2000): "Understanding why farmers plant trees in the homestead agroforestry in Bangladesh." Agroforestry Systems, 50(1), 77-93.

Sanchez, P. (1995): "Science in agroforestry." Agroforestry Systems, 30(1 - 2), 5-55.

Sanchez, P. A., Buresh, R. J., and Leakey, R. R. B. (1997): "Trees, soils, and food security." Philosophical Transactions of the Royal Society of London Series B-Biological Sciences, 352(1356), 949-960.

Sands, R. D. (2004): "Dynamics of carbon abatement in the Second Generation Model." Energy Economics, 26(4), 721-738.

Saris, W. E., Satorra, A., and Sörbom, D. (1987): "The detection and correction of specification errors in structural equation models." Sociological Methodology 1987, C. Clogg, ed., Jossey-Bass, San Francisco, 105-130.

Satorra, A. and Saris, W. E. (1985): "Power of the Likelihood Ratio Test in Covariance Structure-Analysis." Psychometrika, 50(1), 83-90.

Scherr, S. J. (1992): "Not Out of the Woods Yet: Challenges for Economics Research on Agroforestry." American Journal of Agricultural Economics, 74(3), 802-808.

Scherr, S. J. (1995a): "Economic-Factors in Farmer Adoption of Agroforestry - Patterns Observed in Western Kenya." World Development, 23(5), 787-804.

Scherr, S. J. (1995b): "Meeting household needs: Farmer tree-growing strategies in western Kenya." Tree Management in Farmer Strategies: Responses to Agricultural Intensification, J. E. M. Arnold and P. A. Dewees, eds., Oxford University Press, Oxford, 141-173.

Schiff, M. (1992): "Social Capital, Labor Mobility, and Welfare: The Impact of Uniting States." Rationality and Society, 4(2), 157-175.

Scholz, S. and Noble, I. (2005): "Generation of Sequestration Credits under the CDM." Legal Aspects of Implementing the Kyoto Protocol Mechanisms: Making Kyoto Work, D. Freestone and C. Streck, eds., Oxford University Press, Oxford, 265-280.

Scholz, S. M. and Jung, M. (2008): "Forestry Projects under the Clean Development Mechanism and Joint Implementation: Rules and Regulations." Climate Change and Forests: Emerging Policy and Market Opportunities, C. Streck, R. O'Sullivan, T. Janson-Smith, and R. Tarasofsky, eds., Chatham House and Brookings Institution Press, London, U.K., Washington D.C., USA, 71-85.

Schulze, E. D., Valentini, R., and Sanz, M. J. (2002): "The long way from Kyoto to Marrakesh: Implications of the Kyoto Protocol negotiations for global ecology." Global Change Biology, 8(6), 505-518.

Schumacker, R. E. and Lomax, R. G. (2004): A Beginner`s Guide to Structural Equation Modeling. Second Edition., Lawrence Erlbaum Associates, Mahwah, New Jersey.

Sharrow, S. H. (1999): "Silvopastoralism: competition and facilitation between trees, livestock and improved grass-clover pastures on temperate rainfed lands." Agroforestry in Sustainable Agricultural Systems, L. E. Buck, J. P. Lassoie, and E. C. M. Fernandes, eds., CRC Press LLC, Boca Raton, Florida, 111-130.

Shively, G. E. (1999): "Prices and tree planting on hillside farms in Palawan." World Development, 27(6), 937-949.

Siegel, S. and Castellan, N. J. (1988): Nonparametric statistics for the behavioral sciences, New York.

Singh, I., Squire, L., and Strauss, J. (1986): Agricultural Household Models: Extensions, Applications, and Policy, The Johns Hopkins University Press, Baltimore, USA and London, UK.

Smith, J. and Scherr, S. J. (2003): "Capturing the value of forest carbon for local livelihoods." World Development, 31(12), 2143-2160.

Staal, S. J., Baltenweck, I., Waithaka, M. M., deWolf, T., and Njoroge, L. (2002): "Location and uptake: integrated household and GIS analysis of technology adoption and land use, with application to smallholder dairy farms in Kenya." Agricultural Economics, 27(3), 295-315.

Steiger, J. H. (1990): "Structural Model Evaluation and Modification - An Interval Estimation Approach." Multivariate Behavioral Research, 25(2), 173-180.

Stern, N. (2006): The Economics of Climate Change. The Stern Review, Cambridge University Press, Cambridge, UK.

Stiglitz, J. E. (2005): Cleaning Up Economic Growth. http://www.project-syndicate.org/commentary/stiglitz59, Accessed 20.07.06.

Streck, C. and Scholz, S. M. (2006): "The role of forests in global climate change: whence we come and where we go." International Affairs, 82(5), 861-879.

Sunderlin, W. D., Angelsen, A., Belcher, B., Burgers, P., Nasi, R., Santoso, L., and Wunder, S. (2005): "Livelihoods, forests, and conservation in developing countries: An overview." World Development, 33(9), 1383-1402.

Tanzania National Bureau of Statistics (2006): "National Agriculture Sample Census 2002/2003." Tanzania National Bureau of Statistics, Tanzania Ministry of Agriculture and Food Security, Tanzania Ministry of Water and Livestock Development and Tanzania Ministry of Cooperatives and Marketing, Dar Es Salaam, forthcoming.

Tanzania Vice Presidence's Office (1999): "Proposed National Action Programme to Combat Desertification." Vice President's Office - United Republic of Tanzania, Dar Es Salaam, Tanzania.

The World Bank (2003): 2003 - World Development Indicators, The World Bank, Washington D.C., USA.

The World Bank (2005): World Development Report 2006 - Equity and Development, The World Bank and Oxford University Press, Washington D.C. - USA.

The World Bank (2007): Forest Carbon Partnership Facility - Concept Note, Washington, DC, USA.

TIST (2003): TIST One Pager. http://www.tist.org/tist/docs/TIST%20One%20Pager%20040318.pdf, accessed 15.12.2006, The International Small Group and Tree Planting Program.

TIST (2004): TIST New Group Information Packet. Revised: February 16, 2004. http://www.tist.org/tist/docs/welcomepack/Feb%2016%202004/TIST%20TZ%20English%20Master%20Welcome%20Packet%20Feb%2016%202004.pdf, accessed 15. 12. 2006 ., The International Small Group and Tree Planting Program, Ukuzaji Maendeleo Endelevu Tanzania, Ltd.

Tschakert, P. (2007): "Environmental services and poverty reduction: Options for smallholders in the Sahel." Agricultural Systems, 94(1), 75-86.

Tucker, L. R. and Lewis, C. (1973): "Reliability Coefficient for Maximum Likelihood Factor-Analysis." Psychometrika, 38(1), 1-10.

Uchida, E., Xu, J., Xu, Z., and Rozelle, S. (2007): "Are the poor benefiting from China's land conservation program?" Environment and Development Economics, 12(4), 593-620.

UN (1997): Kyoto Protocol to the United Nations Framework Convention on Climate Change, United Nations; http://unfccc.int/resource/docs/convkp/kpeng.pdf; Accessed 2.10.2006.

UN (2004): "Committee for Development Policy. Report on the sixth session." Official Records 2004, Supplement No. 13, E/2004/33.

UNESCO (2004): Education for all. The quality imperative. Global Monitoring Report 2005, UNESCO - United Nations Educational, Scientific and Cultural Organization, Paris, France.

UNFCCC (1992): United Nations Framework Convention on Climate Change, FCCC/INFORMAL/84; http://unfccc.int/resource/docs/convkp/conveng.pdf; Accessed 2.10.2006.

UNFCCC (2001): "Decision 17/CP.7 - Modalities and procedures for a clean development mechanism as defined in Article 12 of the Kyoto Protocol." FCCC/CP/2001/13/Add. 2.

UNFCCC (2002): "Decision 15/CP.7 - Principles, nature and scope of the mechanisms pursuant Articles 6, 12 and 17 of the Kyoto Protocol." FCCC/CP/2001/13/Add. 2.

UNFCCC (2005a): "Caring for Climate. A guide to the Climate Change Convention and the Kyoto Protocol." J. Depledge and R. Lamb, eds., Climate Change Secretariat (UNFCCC), Bonn, Germany.

UNFCCC (2005b): "Decision 5/CMP.1 - Modalities and procedures for afforestation and reforestation project activities under the clean development mechanism in the first commitment period of the Kyoto Protocol." FCCC/KP/CMP/2005/8/Add. 1.

UNFCCC (2005c): "EB 21 Report Annex 16 - Tool for the Demonstration and Assessment of Additionality in A/R CDM Project Activities." http://cdm.unfccc.int/methodologies/ARmethodologies/AdditionalityTools/Additionality_tool.pdf, Accessed 20. 07. 2006.

UNFCCC (2006a): Background paper for the workshop on reducing emissions from deforestation in developing countries, Framework Convention on Climate Change Secretariat, Bonn, Germany.

UNFCCC (2006b): CDM project activity cycle. http://cdm.unfccc.int/CommonImages/ProjectCycleSlide, Accessed 3. 10. 2006 .

UNFCCC (2006c): "Decision 16/CMP.1 - Land use, land-use change and forestry." FCCC/KP/CMP/2005/8/Add. 3.

UNFCCC (2006d): "Decision 5/CMP.1 - Modalities and procedures for afforestation and reforestation project activities under the clean development mechanism in the first commitment period of the Kyoto Protocol." FCCC/KP/CMP/2005/8/Add. 1.

URT (2006a): Dodoma Region. http://www.tanzania.go.tz/regions/dodoma/index.html.

URT (2006b): The United Republic of Tanzania, National Website, Agriculture. http://www.tanzania.go.tz/agriculturef.html., URT - The United Republic of Tanzania.

Watson, R. T. and Noble, I. R. (2004): The global imperative and policy for carbon sequestration: The Carbon Balance of Forest Biomes, Garland Science/BIOS Scientific Publishers, New York, USA.

Welch, F. (1978): "The role of investment in human capital in agriculture." Distortion of Agricultural Incentives, T. W. Schulz, ed., Indiana University Press, Bloomington, USA.

West, S. G., Finch, J. F., and Curran, P. J. (1995): "Structural equation models with nonnormal variables: Problems and remedies." Structural equation modeling: Concepts, issues, and applications, R. H. Hoyle, ed., Sage, Thousand Oaks, CA, 56-75.

Wheaton, B., Muthén, B., Alwin, D. F., and Summers, G. F. (1977): "Assessing reliability and stability in panel models." Sociological methodology, D. R. Heise, ed., Jossey-Bass, San Francisco, 84-136.

WHO (2006): "The World Health Report 2006: working together for health." World Health Organization - WHO Press, Geneva, Switzerland.

Williams, L. J. and Holahan, P. J. (1994): "Parsimony-based fit indices of multiple indicator models: Do they work?" Structural Equation Modeling: A Multidisciplinary Journal, 1, 161-189.

Woolcock, M. (1998): "Social capital and economic development: Toward a theoretical synthesis and policy framework." Theory and Society, 27(2), 151-208.

Wright, S. (1918): "On the nature of size factors." Genetics, 3, 367-374.

Wright, S. (1921): "Correlation and causation." Journal of Agricultural Research, 20, 557-585.

Wunder, S. (2005): "Payments for environmental services: Some nuts and bolts." CIFOR - Center for International Forestry Research, Bogor, Indonesia.

Wunder, S. (2007): "The efficiency of payments for environmental services in tropical conservation." Conservation Biology, 21(1), 48-58.

Xu, Z., Bennett, M. T., Tao, R., and Xu, J. (2004): "Chinas`s Sloping Land Conversion Programme four years on: current situation and pending issues." International Forestry Review, 6(3-4), 317-326.

Yang, R. Y., Tsou, S. C. S., Lee, T. C., Chang, L. C., Kuo, G., and Lai, P. Y. (2006): "Moringa, a novel plant rich in antioxidants, bioavailable iron, and nutrients." Herbs: Challenges in Chemistry and Biology, 925, 224-239.

Zablon, J. (2005): personal communication: Sisal (Agave sisalana Perrine) and its usage in the rural Tanzanian household.

Zbinden, S. and Lee, D. R. (2005): "Paying for Environmental Services: An Analysis Of Participation in Costa Rica's PSA Program." World Development, 33(2), 255-272.

Development Economics and Policy

Series edited by Franz Heidhues, Joachim von Braun and Manfred Zeller

Band 1 Andrea Fadani: Agricultural Price Policy and Export and Food Production in Cameroon. A Farming Systems Analysis of Pricing Policies. The Case of Coffee-Based Farming Systems. 1999.

Band 2 Heike Michelsen: Auswirkungen der Währungsunion auf den Strukturanpassungsprozeß der Länder der afrikanischen Franc-Zone. 1995.

Band 3 Stephan Bea: Direktinvestitionen in Entwicklungsländern. Auswirkungen von Stabilisierungsmaßnahmen und Strukturreformen in Mexiko. 1995.

Band 4 Franz Heidhues / François Kamajou: Agricultural Policy Analysis – Proceedings of an International Seminar, held at the University of Dschang, Cameroon on May 26 and 27 1994, funded by the European Union under the Science and Technology Program (STD). 1996.

Band 5 Elke M. Förster: Protection or Liberalization? A Policy Analysis of the Korean Beef Sector. 1996.

Band 6 Gertrud Schrieder: The Role of Rural Finance for Food Security of the Poor in Cameroon. 1996.

Band 7 Nestor R. Ahoyo Adjovi: Economie des Systèmes de Production intégrant la Culture de Riz au Sud du Bénin: Potentialités, Contraintes et Perspectives. 1996.

Band 8 Jenny Müller: Income Distribution in the Agricultural Sector of Thailand. Empirical Analysis and Policy Options. 1996.

Band 9 Michael Brüntrup: Agricultural Price Policy and its Impact on Production, Income, Employment and the Adoption of Innovations. A Farming Systems Based Analysis of Cotton Policy in Northern Benin. 1997.

Band 10 Justin Bomda: Déterminants de l'Epargne et du Crédit, et leurs Implications pour le Développement du Système Financier Rural au Cameroun. 1998.

Band 11 John M. Msuya: Nutrition Improvement Projects in Tanzania: Implementation, Determinants of Performance, and Policy Implications. 1998.

Band 12 Andreas Neef: Auswirkungen von Bodenrechtswandel auf Ressourcennutzung und wirtschaftliches Verhalten von Kleinbauern in Niger und Benin. 1999.

Band 13 Susanna Wolf (ed.): The Future of EU-ACP Relations. 1999.

Band 14 Franz Heidhues / Gertrud Schrieder (eds.): Romania – Rural Finance in Transition Economies. 2000.

Band 15 Katinka Weinberger: Women's Participation. An Economic Analysis in Rural Chad and Pakistan. 2000.

Band 16 Christof Batzlen: Migration and Economic Development. Remittances and Investments in South Asia: A Case Study of Pakistan. 2000.

Band 17 Matin Qaim: Potential Impacts of Crop Biotechnology in Developing Countries. 2000.

Band 18 Jean Sènahoun: Programmes d'ajustement structurel, sécurité alimentaire et durabilité agricole. Une approche d'analyse intégrée, appliquée au Bénin. 2001.

Band 19 Torsten Feldbrügge: Economics of Emergency Relief Management in Developing Countries. With Case Studies on Food Relief in Angola and Mozambique. 2001.

Band 20 Claudia Ringler: Optimal Allocation and Use of Water Resources in the Mekong River Basin: Multi-Country and Intersectoral Analyses. 2001.

Band 21 Arnim Kuhn: Handelskosten und regionale (Des-)Integration. Russlands Agrarmärkte in der Transformation. 2001.

Band 22 Ortrun Anne Gronski: Stock Markets and Economic Growth. Evidence from South Africa. 2001.

Band 23 Patrick Webb / Katinka Weinberger (eds.): Women Farmers. Enhancing Rights, Recognition and Productivity. 2001.

Band 24 Mingzhi Sheng: Lebensmittelkonsum und -konsumtrends in China. Eine empirische Analyse auf der Basis ökonometrischer Nachfragemodelle. 2002.

Band 25 Maria Iskandarani: Economics of Household Water Security in Jordan. 2002.

Band 26 Romeo Bertolini: Telecommunication Services in Sub-Saharan Africa. An Analysis of Access and Use in the Southern Volta Region in Ghana. 2002.

Band 27 Dietrich Müller-Falcke: Use and Impact of Information and Communication Technologies in Developing Countries' Small Businesses. Evidence from Indian Small Scale Industry. 2002.

Band 28 Wolfram Erhardt: Financial Markets for Small Enterprises in Urban and Rural Northern Thailand. Empirical Analysis on the Demand for and Supply of Financial Services, with Particular Emphasis on the Determinants of Credit Access and Borrower Transaction Costs. 2002.

Band 29 Wensheng Wang: The Impact of Information and Communication Technologies on Farm Households in China. 2002.

Band 30 Shyamal K. Chowdhury: Institutional and Welfare Aspects of the Provision and Use of Information and Communication Technologies in the Rural Areas of Bangladesh and Peru. 2002.

Band 31 Annette Luibrand: Transition in Vietnam. Impact of the Rural Reform Process on an Ethnic Minority. 2002.

Band 32 Felix Ankomah Asante: Economic Analysis of Decentralisation in Rural Ghana. 2003.

Band 33 Chodechai Suwanaporn: Determinants of Bank Lending in Thailand: An Empirical Examination for the Years 1992 to 1996. 2003.

Band 34 Abay Asfaw: Costs of Illness, Demand for Medical Care, and the Prospect of Community Health Insurance Schemes in the Rural Areas of Ethiopia. 2003.

Band 35 Gi-Soon Song: The Impact of Information and Communication Technologies (ICTs) on Rural Households. A Holistic Approach Applied to the Case of Lao People's Democratic Republic. 2003.

Band 36 Daniela Lohlein: An Economic Analysis of Public Good Provision in Rural Russia. The Case of Education and Health Care. 2003.

Band 37 Johannes Woelcke. Bio-Economics of Sustainable Land Management in Uganda. 2003.

Band 38 Susanne M. Ziemek: The Economics of Volunteer Labor Supply. An Application to Countries of a Different Development Level. 2003.

Band 39 Doris Wiesmann: An International Nutrition Index. Concept and Analyses of Food Insecurity and Undernutrition at Country Levels. 2004.

Band 40 Isaac Osei-Akoto: The Economics of Rural Health Insurance. The Effects of Formal and Informal Risk-Sharing Schemes in Ghana. 2004.

Band 41 Yuansheng Jiang: Health Insurance Demand and Health Risk Management in Rural China. 2004.

Band 42 Roukayatou Zimmermann: Biotechnology and Value-added Traits in Food Crops: Relevance for Developing Countries and Economic Analyses. 2004.

Band 43 F. Markus Kaiser: Incentives in Community-based Health Insurance Schemes. 2004.

Band 44 Thomas Herzfeld: *Corruption begets Corruption.* Zur Dynamik und Persistenz der Korruption. 2004.

Band 45 Edilegnaw Wale Zegeye: The Economics of On-Farm Conservation of Crop Diversity in Ethiopia: Incentives, Attribute Preferences and Opportunity Costs of Maintaining Local Varieties of Crops. 2004.

Band 46 Adama Konseiga: Regional Integration Beyond the Traditional Trade Benefits: Labor Mobility contribution. The Case of Burkina Faso and Côte d'Ivoire. 2005.

Band 47 Beyene Tadesse Ferenji: The Impact of Policy Reform and Institutional Transformation on Agricultural Performance. An Economic Study of Ethiopian Agriculture. 2005.

Band 48 Sabine Daude: Agricultural Trade Liberalization in the WTO and Its Poverty Implications. A Study of Rural Households in Northern Vietnam. 2005.

Band 49 Kadir Osman Gyasi: Determinants of Success of Collective Action on Local Commons. An Empirical Analysis of Community-Based Irrigation Management in Northern Ghana. 2005.

Band 50 Borbala E. Balint: Determinants of Commercial Orientation and Sustainability of Agricultural Production of the Individual Farms in Romania. 2006.

Band 51 Pamela Marinda: Effects of Gender Inequality in Resource Ownership and Access on Household Welfare and Food Security in Kenya. A Case Study of West Pokot District. 2006.

Band 52 Charles Palmer: The Outcomes and their Determinants from Community-Company Contracting over Forest Use in Post-Decentralization Indonesia. 2006.

Band 53 Hardwick Tchale: Agricultural Policy and Soil Fertility Management in the Maize-based Smallholder Farming System in Malawi. 2006.

Band 54 John Kedi Mduma: Rural Off-Farm Employment and its Effects on Adoption of Labor Intensive Soil Conserving Measures in Tanzania. 2006.

Band 55 Mareike Meyn: The Impact of EU Free Trade Agreements on Economic Development and Regional Integration in Southern Africa. The Example of EU-SACU Trade Relations. 2006.

Band 56 Clemens Breisinger: Modelling Infrastructure Investments, Growth and Poverty Impact. A Two-Region Computable General Equilibrium Perspective on Vietnam. 2006.

Band 57 Meike Wollni: Coping with the Coffee Crisis. An Analysis of the Production and Marketing Performance of Coffee Farmers in Costa Rica. 2007.

Band 58 Franklin Simtowe: Performance and Impact of Microfinance. Evidence from Joint Liability Lending Programs in Malawi. 2008.

Band 59 Xiangping Jia: Credit Rationing and Institutional Constraint. Evidence from Rural China. 2008.

Band 60 Holger Seebens: The Economics of Gender and the Household in Developing Countries. 2009.

Band 61 Stephan Piotrowski: Land Property Rights and Natural Resource Use. An Analysis of Household Behavior in Rural China. 2009.

Band 62 Sebastian M. Scholz: Rural Development through Carbon Finance. Forestry Projects under the Clean Development Mechanism of the Kyoto Protocol. Assessing Smallholder Participation by Structural Equation Modeling. 2009.

Band 63 Jakob Rupert Friederichsen: Opening Up Knowledge Production through Participatory Research? Agricultural Research for Vietnam's Northern Uplands. 2009.

www.peterlang.de

Peter Lang · Internationaler Verlag der Wissenschaften

John Kedi Mduma

Rural Off-Farm Employment and its Effects on Adoption of Labor Intensive Soil Conserving Measures in Tanzania

Frankfurt am Main, Berlin, Bern, Bruxelles, New York, Oxford, Wien, 2006.
XV, 135 pp., 13 fig., 14 tab.
Development economics and policy.
Series edited by Franz Heidhues and Joachim von Braun. Vol. 54
ISBN 978-3-631-54652-9 · pb. € 36.20*

This study analyzes two related aspects: households' participation in off-farm employment as a livelihood strategy and the effects of participation in off-farm employment on households' adoption of labor intensive soil conserving technologies. Several factors are significant but spatial econometric analysis reveals that model parameters vary substantially across space. In addition, households supplying labor off-farm are generally associated with reduced adoption of terraces, hedgerows and cut-offs. The negative impact of supplying labor off-farm can be moderately cushioned when households also hire labor to work on the construction or maintenance of soil conserving structures. However, it is shown that hired labor is not a perfect substitute for households' own labor and does not fully off-set the effect of a household's off-farm labor supply.

Contents: The Agricultural Sector and Rural Households in Tanzania · Farm Household Model: A Theoretical Framework · Households' Off-Farm Labor Allocation Decisions · Spatial Analysis of Village Level Labor Markets · Off-Farm Employment and Household's Decision on Soil Conservation Technologies

Frankfurt am Main · Berlin · Bern · Bruxelles · New York · Oxford · Wien
Distribution: Verlag Peter Lang AG
Moosstr. 1, CH-2542 Pieterlen
Telefax 00 41 (0) 32 / 376 17 27

*The €-price includes German tax rate
Prices are subject to change without notice

Homepage http://www.peterlang.de

www.ingramcontent.com/pod-product-compliance
Ingram Content Group UK Ltd.
Pitfield, Milton Keynes, MK11 3LW, UK
UKHW021842100926
8556IPUK00019B/1000

9 783631 592502